Biological Wastewater Treatment Process Design Calculations

Harlan H. Bengtson, PE, PhD

Emeritus Professor of Civil Engineering
Southern Illinois University Edwardsville

Copyright © 2017 by Harlan H. Bengtson

Table of Contents

	Page
Introduction …………………………………………..	2
Chapter 1 – Biochemical Oxygen Demand as a cause of Water Pollution ……	4
Chapter 2 – Description of the Activated Sludge Process ……...	7
Chapter 3 – Activated Sludge Process Variations …………….	9
Chapter 4 – Activated Sludge Parameters ……………………	13
Chapter 5 – Activated Sludge Process Design Calculations ……	15
Chapter 6 – Activated Sludge Operational Calculations ……….	21
Chapter 7 – Oxygen/Air/Blower Calculations and Alkalinity Requirements ……	29
Chapter 8 – Description of the MBBR (Moving Bed Biofilm Reactor) ……	33
Chapter 9 – Single Stage BOD Process Design Calculations ……	38
Chapter 10 - Two-Stage BOD Removal MBBR Process Design Calculations ……	47
Chapter 11 – Single Stage Nitrification MBBR Process Design Calculations ……	53
Chapter 12 – Two Stage BOD Removal Nitrification Process Design Calculations ……	61
Chapter 13 – Denitrification Background Information …………	69
Chapter 14 – Post-Anoxic Denitrification Process Design Calculations …….	72
Chapter 15 – Pre-Anoxic Denitrification Process Design Calculations …….	80
Chapter 16 – Description of the MBR (Membrane Bioreactor) Process ………	87
Chapter 17 – Membrane Module Process Design Calculation ….	90
Chapter 18 – Process Design Calculations for MBR BOD Removal/Nitrification ……	93
Chapter 19 – Process Design Calculations for MBR Pre-Anoxic Denitrification ……	107
References ……………………………………………….	114

Introduction

Biological wastewater treatment is very widely used for removal of biodegradable materials from wastewater. This book starts with a discussion of the biochemical oxygen demand that is created by biodegradable materials in water and the reason why such materials must be removed from wastewater.

The three biological wastewater treatment processes that will be discussed in the book are activated sludge, MBBR (moving bed biofilm reactor), and MBR (membrane bioreactor). The activated sludge process, developed in 1914 in England, is definitely the "old timer" of this group. The MBR process first appeared in the 1970s, but it was not until the use of a membrane module immersed in the aeration tank was introduced in 1989, that its use became more widespread. The MBBR process was developed in Norway in the 1990s and its use has spread quite rapidly since then.

Additional description and background information about each of these processes will be provided along with a presentation of a process design calculation procedure and illustrative example calculations for each of them.

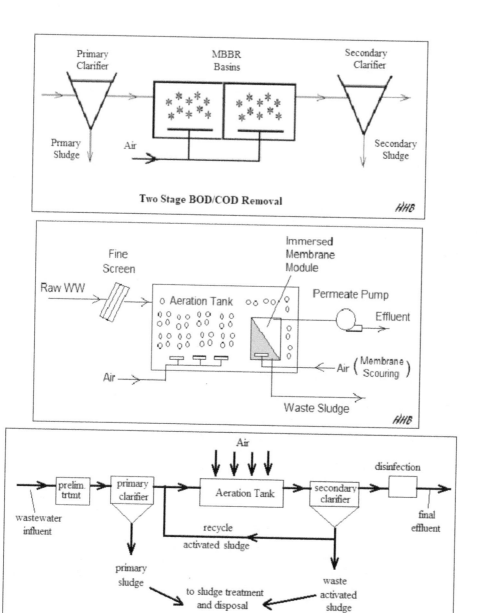

Figure 1. Flow Diagrams for Activated Sludge, MBBR, and MBR Processes

Biochemical Oxygen Demand as a Cause of Water Pollution

Biochemical oxygen demand (BOD) is an indirect measure of the concentration of biodegradable organic matter in water or wastewater. Organic matter (as measured by BOD) is one of the major constituents removed from wastewater in domestic wastewater treatment plants. The reason for being concerned about organic matter in water is its effect on dissolved oxygen in the receiving stream. Dissolved oxygen in water is essential for much of aquatic life, so organic contaminants that affect dissolved oxygen level in water are of concern. The "death and decay" portion of the organic carbon cycle shown in the **Figure 2** diagram below is the portion that takes place in the biological treatment component of a wastewater treatment plant or else takes place in the receiving stream if the organic matter isn't removed in the treatment plant.

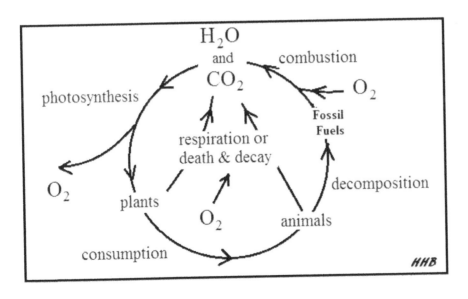

Figure 2. The Organic Carbon Cycle

The two major reactions that take place in the organic carbon cycle are biological oxidation of waste organic matter and photosynthesis, which is the process by which green plants produce organic matter from carbon dioxide and water in reactions that are catalyzed by sunlight and the chlorophyll in the green plants. Through the biological oxidation process, aerobic microorganisms utilize oxygen in breaking down organic matter to carbon dioxide and water together with small amounts of other end products.

The photosynthesis and biological oxidation processes can be represented by the following two equations:

Photosynthesis:

CO_2 + H_2O + sunlight → **organic plant matter** (primarily C, H, & O) + **oxygen** (this reaction is catalyzed by the chlorophyll in green plants)

Biological Oxidation:

waste organic matter (primarily C, H & O) + **O_2** → **CO_2 + H_2O + energy**

This reaction is the 'death and decay' shown in the organic carbon cycle diagram. The process takes place as aerobic microorganisms utilize the waste organic matter as their food (energy) source. The process uses oxygen, so if it is taking place in a water body, dissolved oxygen is consumed. A large quantity of organic matter in the water will result in multiplication of microorganisms and rapid removal of dissolved oxygen, leading to oxygen depletion below the level needed by aquatic life.

This is also the process that takes place in biological oxidation processes in wastewater treatment plants for removal of organic matter from the incoming wastewater.

Description of the Activated Sludge Process

The activated sludge process is very widely used for biological wastewater treatment. **Figure 3** below shows a general flow diagram with the typical components present in an activated sludge wastewater treatment plant.

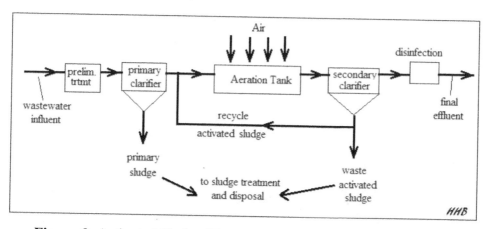

Figure 3. Activated Sludge Wastewater Treatment Flow Diagram

The first component is preliminary treatment, typically consisting of screening, flow measurement, and perhaps grit removal. The second component, the primary clarifier, is used to remove settleable suspended matter. The underflow goes to sludge treatment and disposal and the overflow goes to an aeration tank. The aeration tank is the heart of an activated sludge treatment process. It is here that biological oxidation of dissolved and fine suspended organic matter takes place. The biological oxidation takes place because aerobic microorganisms, organic matter and dissolved oxygen are all brought together in the aeration tank. The organic matter comes in with the primary effluent. The dissolved oxygen level is maintained by blowing air into the aeration tank through diffusers (or in some cases with a mechanical aerator). This also serves to keep the aeration tank contents mixed. A suitable concentration of microorganisms is maintained in the aeration tank by settling out the 'activated sludge' (microorganisms) in the secondary clarifier and recycling them back into the aeration tank.

The activated sludge process was first developed in England in the early 1900s by E. Ardern and W.T. Lockett. They presented their findings to the Manchester section meeting of the Society of Chemical Industry on April 3, 1914. The essentials of the activated sludge process, as described by Ardern and Lockett are:

1. Aeration of wastewater in the presence of aerobic microorganisms
2. Removal of biological solids from the wastewater by sedimentation
3. Recycling of the settled biological solids back into the aerated wastewater

These three components, shown in the flow diagram above, are still the essence of the activated sludge process, as it is used today.

Activated Sludge Process Variations

Four common variations of the activated sludge process are:

1. Conventional activated sludge
2. Extended aeration
3. Completely mixed activated sludge
4. The contact stabilization process

A brief description of each follows.

The Conventional Activated Sludge Process is used over a wide range of wastewater flow rates, from small to very large plants. The flow diagram and general description is that given above in the Activated Sludge Background section. The aeration tank in a conventional activated sludge process is typically designed with a long, narrow configuration, thus giving approximately 'plug flow' through the tank. For large treatment plants, the aeration tank is often built with a serpentine pattern, like that shown in **Figure 4** below, in order to obtain the desired plug flow without an excessive length requirement for the tank.

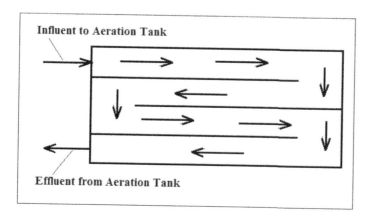

Figure 4. Serpentine Pattern, Plug Flow Aeration Tank

The Extended Aeration Activated Sludge Process is shown in **Figure 5** below. As you can see, this process doesn't use a primary clarifier. Instead, a longer detention time is used for the aeration tank, so that the settleable organic matter will be biologically oxidized along with the dissolved and fine suspended organic matter. This requires a hydraulic detention time of about 24 hours instead of the 6 to 8 hours that is typical for the conventional activated sludge process. This simplifies the operation of the plant by eliminating the primary clarifier and reducing the need for sludge treatment and disposal to a very minimal flow of waste activated sludge that must be drawn off periodically.

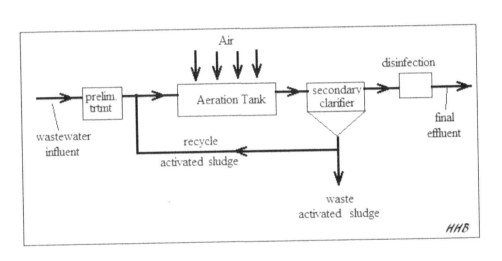

Figure 5. Extended Aeration Activated Sludge Process Flow Diagram

The Completely Mixed Activated Sludge Process has the same overall flow pattern as the conventional activated sludge process. The main differences are the method of aeration and the aeration tank configuration. For the completely mixed option, aeration is usually with a mechanical mixer, rather than with diffused air. Also, the tank configuration is usually approximately square, rather than long and narrow. This combination of mixing and tank configurations makes the aeration tank approximate a completely mixed reactor rather than the plug flow reactor approximated by the conventional activated sludge aeration tank. The flow diagram below illustrates this.

Typical applications of the completely mixed activated sludge process are cases where slug flows of high concentration, hard to oxidize, or toxic wastes enter the treatment plant. The complete mixing dilutes such flows into the entire tank contents more rapidly than a plug flow design, making the slug flow less likely to upset or kill the microorganisms.

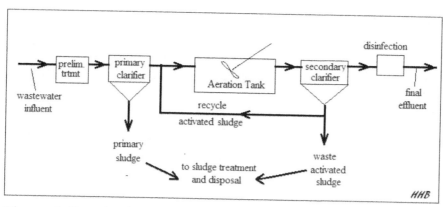

Figure 6. Completely Mixed Activated Sludge Process Flow Diagram

<u>The Contact Stabilization Activated Sludge Process</u> gets by with less total aeration tank volume than that needed for the conventional activated sludge process. This is accomplished because the full wastewater flow is aerated for only 0.5 to 2 hours in an aerated contact tank. This is sufficient time for removal of the organic matter from the wastewater flow by the microorganism. If those microorganisms were recycled directly into the aeration tank after settling out in the secondary clarifier, however,

they would not continue to take up organic matter, because they are "still full" from the 0.5 to 2 hour feast they recently had. If the recycle sludge is aerated for 3 to 8 hours to allow the microorganisms to "digest" the organic matter that they've taken up, then they go back into the aeration tank ready to go to work. Since the recycle activated sludge flow is less than the full wastewater flow, this results in less overall aeration tank volume for a given wastewater flow rate to be treated. The diagram below shows this process.

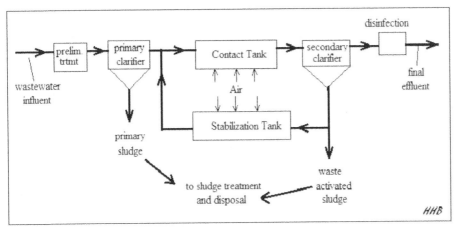

Figure 7. Contact Stabilization Activated Sludge Process Flow Diagram

Activated Sludge Parameters

Figure 8 below shows an activated sludge aeration tank and secondary clarifier with parameters for the primary effluent, secondary effluent, waste activated sludge, and recycle activated sludge streams.

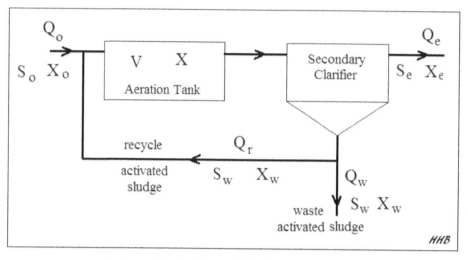

Figure 8. Activated Sludge Parameters

The parameters in the diagram and a few others that will be used for the upcoming activated sludge calculations are summarized in the list below.

- primary effluent flow rate, Q_o, MGD (m³/day for S.I.)

- primary effluent biochemical oxygen demand (BOD) concentration, S_o, mg/L (g/m³ for S.I.)

- primary effluent suspended solids conc., X_o, mg/L (g/m³ for S.I.)

- aeration tank volume, V, ft³ (m³ for S.I.)

- aeration tank MLSS (suspended solids conc.), **X**, mg/L (g/m^3 for S.I.)

- secondary effluent flow rate, **Q_e**, MGD, (m^3/day for S.I.)

- secondary effluent susp.solids conc., **X_e**, mg/L (g/m^3 for S.I.)

- secondary effluent biochemical oxygen demand (BOD) concentration, **S_e**, mg/L (g/m^3 for S.I.)

- waste activated sludge flow rate, **Q_w**, MGD (m^3/day for S.I.)

- waste activated sludge biochemical oxygen demand (BOD) conc., **S_w**, mg/L (g/m^3 for S.I.)

- waste activated sludge susp. solids conc., **X_w** mg/L (g/m^3 for S.I.)

- recycle activated sludge flow rate, **Q_r**, MGD (m^3/day for S.I.)

- Food to Microorganism ratio, **F:M**, lb BOD/day/lb MLVSS (kg BOD/day/kg MLVSS)

- Hydraulic retention time, **HRT**, hours (hours for S.I.)

- Sludge retention time (also called sludge age), **SRT**, days (days for S.I.)

- Volumetric loading, **VL**, lb BOD/day/1000 ft^3 (kg BOD/day/m^3 for S.I.)

- % volatile solids in the aeration tank mixed liquor suspended solids, **%Vol**.

Activated Sludge Process Design Calculations

Table 1 below shows typical values for three design parameters sometimes used for sizing activated sludge aeration basins: volumetric loading, food to microorganism ratio (F:M), and hydraulic residence time (HRT). Note that values for volumetric loading are given in both U.S. and S.I. units. The other two parameters are the same for either U.S. or S.I. units, since F:M will be the same expressed as either lb BOD/day/lb MLVSS or kg BOD/day/kg MLVSS, and HRT is simply in hours. **Table 1** is adapted from information in Metcalf & Eddy (Ref #2).

Table 1. Activated Sludge Aeration Tank Design Parameters – Typical Values

Activated Sludge Process	Volumetric Loading lb BOD/day 1000 ft^3	Volumetric Loading kg BOD/day m^3	F/M kg BOD/day kg MLSS	HRT hours
Conventional Plug Flow	20 - 40	0.3 - 0.7	0.2 - 0.4	4 - 8
Complete Mix	20 - 100	0.3 - 1.6	0.2 - 0.6	3 - 5
Extended Aeration	5 - 15	0.1 - 0.3	0.04 - 0.1	20 - 30

NOTE: F:M values will be the same for units of lb BOD/day/lb MLVSS.

Calculations with these design parameters can be made in U.S. units using the following equations:

- $V = [(8.34 \cdot S_o \cdot Q_o)/V_L](1000)$

- $V_{MG} = V \cdot 7.48/1{,}000{,}000$

- $HRT = 24 \cdot V_{MG}/Q_o$

- $F:M = (8.34 \cdot S_o \cdot Q_o)/(8.34 \cdot \%Vol \cdot X \cdot V_{MG})$

  $= (S_o \cdot Q_o)/(\%Vol \cdot X \cdot V_{MG})$

V_{MG} is the tank volume in millions of gallons. It is introduced for convenience in calculations, since the primary effluent flow rate is given in MGD. The other parameters in the equations are as defined in the list above. The 8.34 factor in the equations above is used to convert mg/L to lb/MG, and the 7.48 is for conversion of ft^3 to gallons. Also, note that the primary sludge flow rate is typically very small in comparison with the influent wastewater flow rate, so the primary effluent flow rate, Q_o, is typically taken to be equal to the plant influent flow rate.

Example #1: Calculate the aeration tank volume requirement for a conventional activated sludge plant treating a daily average flow rate of 3.5 MGD, with primary effluent BOD estimated to be 175 mg/L. The design criterion is to be a volumetric loading rate of 30 lb BOD/day/1000 ft^3.

Solution: The required volume can be calculated from the first equation in the list above:

$V = [(8.34 \cdot S_o \cdot Q_o)/V_L](1000) = 8.34 \cdot 175 \cdot 3.5 \cdot 1000/30 = \underline{\mathbf{170{,}275\ ft^3}}$

Example #2: For an assumed aeration tank MLSS of 2100 mg/L and assumed % volatile MLSS of 75%, what would be the aeration tank F:M ratio and hydraulic residence time for the plant inflow and aeration tank volume from Example #1?

Solution: The HRT and F:M ratio can be calculated using the last three equations in the list above, as follows:

$V_{MG} = V*7.48/1,000,000 = 170,275*7.48/1,000,000 = 1.27$ MG

$HRT = 24*V_{MG}/Q_o = 24*1.27/3.5 =$ **8.7 hours**

$F:M = (8.34*S_o*Q_o)/(8.34*\%Vol*X*V_{MG}) =$
$(175*3.5)/(0.75*2100*1.27)$

= **0.31 lb BOD/day/lb MLVSS**

Example #3: Set up and use an Excel spreadsheet to find the solutions to Examples #1 and #2.

Solution: The screenshot in **Figure 9** on the next page shows an Excel spreadsheet with the solution to this example.

This spreadsheet is set up for user entry of specified design volumetric loading, **VL**, and input values for primary effluent flow rate, Q_o, primary effluent biochemical oxygen demand (BOD), S_o, aeration tank MLSS, **X**, and % volatile solids in the aeration tank, **%Vol**.

The spreadsheet then uses the equations presented above to calculate the design aeration tank volume and the resulting values for the other two design parameters, **F:M** and **HRT**.

To solve this Example problem, the given values (Q_o = 3.5 MGD, S_o = 175 mg/L, **VL** = 30   lb BOD/day/1000 ft^3,   **X** = 2100 mg/L,   **% Vol** = 75%) were entered into the blue cells in the upper left portion of the spreadsheet. The spreadsheet then calculated the parameters in the yellow cells in the upper right portion of the spreadsheet. As shown in the screenshot below, the same values for **V, HRT,** and **F:M** are obtained as those shown above for Example #1 and Example #2:

V = 170,275 ft^3   **HRT = 8.7 hours**

F:M = 0.31 lb BOD/day/lb MLVSS

Activated Sludge Waste Water Treatment Calculations - U.S. units						
1. Aeration Tank Design						
Instructions: Enter values in blue boxes. Spreadsheet calculates values in yellow boxes						
Inputs			**Calculations**			
Prim. Effl. Flow Rate, Q_o =	3.5	MGD	(Design Based on Volumetric Loading)			
Prim. Effl. BOD, S_o =	175	mg/L	Aeration tank volume, V =	170,275		ft³
Aeration tank MLSS, X =	2100	mg/L	Aeration tank vol. V_{MG} =	1.27		MG
Design Vol. Loading, VL = (lb BOD/day/1000 ft³)	30		Check on other design parameters:			
			Aeration tank HRT =	8.7		hr
% volatile MLSS, %Vol =	75%					
			Aeration tank $F{:}M$ = (lb BOD/day/lb MLVSS)	0.31		

Figure 9. Screenshot of Spreadsheet Solution to Example #3

Source of spreadsheet for screenshot:
www.EngineeringExcelTemplates.com

Calculations in S.I. units can be made using the following equations:

- $V = (S_o * Q_o / 1000)/VL$

- $HRT = 24 * V / Q_o$

- $F{:}M = (S_o * Q_o)/(\%Vol * X * V)$

The equations are slightly simpler, because the S.I system doesn't have a strange volume unit like the gallon! The S.I. units for all of the parameters are given in the long list above. Note that the S.I. concentration unit g/m^3 is numerically equal to mg/L.

Example #4: Calculate the aeration tank volume requirement for a conventional activated sludge plant treating a daily average flow rate of 20,000 m^3/day, with primary effluent BOD estimated to be 140 g/m^3. The design criterion is to be a volumetric loading rate of 0.5 kg BOD/day/m^3.

Solution: The required volume can be calculated from the first equation in the list above:

$V = (S_o*Q_o/1000)/V_L = (140*20,000/1000)/0.5 =$ **5,600 m^3**

Example #5: For an assumed aeration tank MLSS of 2100 mg/L and assumed % volatile MLSS of 75%, what would be the aeration tank F:M ratio and hydraulic residence time for the plant inflow and aeration tank volume from Example #3?

Solution: The HRT and F:M ratio can be calculated using the last two equations in the list above, as follows:

$HRT = 24*V/Q_o = 24*5,600/20,000 =$ **6.7 hr**

$F:M = (S_o*Q_o)/(\%Vol*X*V) = (140*20,000)/(0.75*2100*5600)$

= **0.32 kg BOD/day/kg MLVSS**

Activated Sludge Operational Calculations

Table 2 below shows typical ranges for several operational activated sludge wastewater treatment process parameters. Note that these values remain the same for U.S. or S.I. units. SRT will still have units of days for the U.S. or S.I. system. MLSS concentration will have the S.I. unit of g/m^3, which is numerically equal to mg/L. F:M will have the S.I. unit of kg BOD/day/kg MLVSS, which is numerically equal to lb BOD/day/lb MLVSS. The % unit for Q_r/Q_o remains the same for U.S. or S.I. units.

Table 2. Activated Sludge Operational Parameters – Typical Values

Activated Sludge Process	SRT days	MLSS mg/L	F/M kg BOD/day kg MLSS	Q_r/Q_o %
Conventional Plug Flow	3 - 15	1000 - 3000	0.2 - 0.4	25 - 75
Complete Mix	3 - 15	1500 - 4000	0.2 - 0.6	25 - 100
Extended Aeration	20 - 40	2000 - 5000	0.04 - 0.1	50 - 150

Calculations with these parameters can be made in U.S. units using the following equations:

- $Q_r = [Q_o(X - X_o) - Q_w X_w]/(X_w - X)$ [sometimes approximated by:

$Q_r = Q_o(X - X_o)/(X_w - X)$] (See discussion in next section of course)

- $V_{MG} = V*7.48/1,000,000$

- $Q_w = (1/X_w)[(V_{MG}*X/SRT) - Q_eX_e]$

- $F:M = (8.34*S_o*Q_o)/(8.34\%Vol*X*V_{MG})$

    $= (S_o*Q_o)/(\%Vol*X*V_{MG})$

The aeration tank volume in millions of gallons, V_{MG}, is used primarily in calculating Q_w and the **F:M** ratio.

Example #6: For the 3.5 MGD activated sludge plant in Example #1, with an aeration tank volume of 170,275 ft³, calculate a) the required recycle activated sludge flow rate (using the simplified equation), b) the waste activated sludge flow rate, and c) the aeration tank F:M ratio, based on the following: primary effluent BOD = 175 mg/L, primary effluent TSS = 200 mg/L, waste/recycle activated sludge SS concentration = 7,000 mg/L, aeration tank MLSS = 2000 mg/L, % volatile solids in the aeration tank = 75%, intended sludge retention time = 12 days.

Solution: a) The sludge recycle rate needed can be calculated from the simplified form of the first equation in the list above:

$Q_r = Q_o(X - X_o)/(X_w - X) = 3.5(2000 - 200)/(7000 - 2000) = \underline{\textbf{1.3 MGD}}$

b) The waste activated sludge flow rate needed to give SRT = 12 days can be calculated from the second and third equations in the list above:

$V_{MG} = V*7.48/1,000,000 = 170,275*7.48/1,000,000 = 1.27$ million gallons

$Q_w = (1/X_w)[(V_{MG}*X/SRT) - Q_eX_e] = (1/7000)[(1.27*2000/12) - (3.5*20)] =$ **0.0203 MGD**

or **20,300 gal/day**

c) The aeration tank F:M ratio can be calculated using the last equation in the list above together with the equation for V_{MG}:

$V_{MG} = V*7.48/1,000,000 = 170,275*7.48/1,000,000 = 1.27$ million gallons

$F:M = (S_o*Q_o)/(\%Vol*X*V_{MG}) = (175*3.5)/(0.75*2000*1.27)$

= **0.321 lb BOD/day/lb MLVSS**

Example #7: Use an Excel spreadsheet to find the solution to Example #6.

Solution: The **Figure 10** screenshot on the next page shows an Excel spreadsheet with the solution to this example.

This spreadsheet is set up for user entry of primary effluent flow rate, Q_o, primary effluent biochemical oxygen demand (BOD), S_o, primary effluent TSS, X_o, recycle/waste activated sludge concentration, X_w, aeration tank volume, **V**, aeration tank MLSS, **X**, % volatile solids in the aeration tank, **%Vol**, and the target value for sludge retention time, **SRT**. The spreadsheet then uses the equations presented above to calculate the recycle activated sludge flow rate, Q_r, the waste activated sludge flow rate, Q_w, and the aeration tank **F:M** ratio.

Activated Sludge Waste Water Treatment Calculations - U.S. units

3. Aeration Tank Operational Calculations

Instructions: Enter values in blue boxes. Spreadsheet calculates values in yellow boxes

Values Transferred from Worksheet 2:

Parameter	Value	Units
Design ww Flow Rate, Q_o =	3.5	MGD
Prim. Effl. BOD, S_o =	175	mg/L
Aeration tank MLSS, X =	2000	mg/L
% volatile MLSS, %Vol =	75%	
Waste/recycle activated sludge SS conc., X_w =	7,000	mg/L
Prim. Effl. TSS, X_o =	200	mg/L
Secondary Effl. TSS, X_e =	20	mg/L

Inputs

Parameter	Value	Units
Aeration tank volume, V =	170,275	ft³
Sludge ret. time, SRT =	12	days

Calculations

Parameter	Value	Units
Aeration tank vol. V_{MG} =	1.27	MG
Aeration tank $F:M$ = (lb BOD/day/lb MLVSS)	0.321	
Waste Act. Sludge Rate, Q_w =	0.0203	MGD
Recycle Activated Sludge Flow Rate, Q_r =	1.26	MGD

Figure 10. Screenshot of Spreadsheet Solution to Example #7

Source of spreadsheet for screenshot:
www.EngineeringExcelTemplates.com

To solve this Example problem, the given values (Q_o = 3.5 MGD, S_o = 175 mg/L, X_o = 200 mg/L, X_w = 7000 mg/L, V = 170,275 ft³, X = 2000 mg/L, % Vol = 75%, SRT = 12 days) were entered into the blue cells or transferred from the previous worksheet. The spreadsheet then calculated the parameters in the cells in the "Calculations" portion of the spreadsheet.

As shown in the screenshot above, the same values for Q_r, Q_w, and **F:M** are obtained as those shown above for Example #1 and Example #2:

$$Q_r = \underline{1.26 \text{ MGD}}$$

$$Q_w = \underline{0.0203 \text{ MGD}}$$

$$F:M = \underline{0.321 \text{ lb BOD/day/lb MLVSS}}$$

Calculations in S.I. units can be made using the following equations:

- $Q_r = [Q_o(X - X_o) - Q_w X_w]/(X_w - X)$ [sometimes approximated by:

  $Q_r = Q_o(X - X_o)/(X_w - X)$] (see discussion in next section of course)

- $Q_w = (1/X_w)[(V*X/SRT) - Q_e X_e]$

- $F:M = (S_o*Q_o)/(\%Vol*X*V)$

More Detail on the Equations for Q_r and Q_w: Note that these equations are the same as those used for U.S. units except that there is no need to calculate V_{MG}, because volume in m³ can be used for all of the calculations.

Most of the equations presented and used above are rather straightforward application of a loading factor, calculation of detention time as volume divided by flow rate, or the equation follows directly from the units. The sources of the equations for Q_w and Q_r aren't quite as obvious, however, so they are discussed briefly here.

Waste Activated Sludge Flow Rate: The equation for waste activated sludge flow rate, Q_w, is based on the principle that the average length of time activated sludge solids stay in the aeration tank [the sludge retention time (SRT) or sludge age] is equal to the mass of solids in the aeration tank divided by the rate at which solids are being wasted from the system. In equation form:

SRT = lb act. sludge in aeration tank/(lb act. sludge leaving system/day)

SRT = $(8.34*X*V_{MG})/[(8.34*X_w*Q_w) + (8.34*X_e Q_e)]$

(Note that the factor 8.34 converts mg/L to lb/MG.)

Units in above equation are: [(lb/MG)*MG]/[(lb/MG)*MG/day]
$\quad$ = lb/(lb/day) = days

solving for Q_w gives the equation in the list above:

$Q_w = (1/X_w)[(V_{MG}*X/SRT) - Q_e X_e]$

<u>Recycle Activated Sludge Flow Rate:</u> An equation for the recycle activated sludge flow rate can be determined by a material balance around the aeration tank. The aeration tank portion of the 'activated sludge parameters' diagram from above is reproduced on the next page. It shows that the inflows to the aeration tank are Q_o with suspended solids concentration of X_o and Q_r with suspended solids concentration of X_w. The outflow from the aeration tank is $Q_o + Q_r$ with suspended solids concentration of X (equal to that in the aeration tank.)

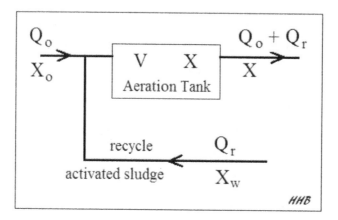

Figure 11. Aeration Tank Material Balance Diagram

A material balance over the aeration tank must take into account the fact that there is a net growth of activated sludge solids in the aeration tank. The material balance is thus:

Rate of solids outflow – Rate of solids inflow = Growth rate of solids

The growth of activated sludge is typically hydraulically controlled with the activated sludge wasting rate and is equal to Q_wX_w. The material balance equation thus becomes:

$$(Q_o + Q_r)X - (Q_oX_o + Q_rX_w) = Q_wX_w$$

For S.I. units, the flow rates in the above equation will be in m³/day and the suspended solids concentrations will be in kg/m³, thus giving units of kg/day for each term in the equation.

For U.S. units, the flow rates in the equation will be in MGD and the suspended solids concentrations will be in mg/L. A factor of 8.34 will be needed with each term to convert mg/L to lb/MG, resulting in lb/day for each term. Since I am going to solve the equation for Q_r, the 8.34 in each term will 'cancel out.'

Through a bit of algebraic manipulation, the equation can be solved for Q_r to give:

$$Q_r = [Q_o(X - X_o) - Q_w X_w]/(X_w - X)$$

The activated sludge wasting rate, Q_w, is typically much less than the influent flow rate, Q_o, so the term $Q_w X_w$ is sometimes dropped out to simplify the equation to:

$$Q_r = Q_o(X - X_o)/(X_w - X)$$

Oxygen/Air/Blower Calculations and Alkalinity Requirements

Oxygen/Air Requirement and Blower Calculations will be made using the "Rule of Thumb" guidelines shown below. These guidelines are made available by Sanitaire, a manufacturer of aeration diffusers.

Rules of Thumb for Estimating Oxygen/Air Requirements - Coarse or fine Bubble Diffusers:

Source: www.xylemwatersolutions.com/scs/sweden/sv-se/produckter/cirkulationpumps/documents/san3.pdf

1. The typical AOR/SOR (or AOTE/SOTE) is 0.50 for a coarse bubble aeration system or 0.33 for a fine bubble aeration system.

2. The typical SOTE is 0.75% per foot (2.46% per m) of diffuser submergence for a coarse bubble system or 2.0% per foot (6.56% per m) of diffuser submergence for a fine bubble system.

3. 1 SCF of air has a density of 0.075 lbm/ft^3 (1.20 $kg/m3$) and contains 23% oxygen by weight, thus: 1 SCF of air contains 0.0173 lbm of oxygen (1SCM contains 0.2770 kg of oxygen).

4. For biological treatment with SRT from 5 to 10 days, mass of oxygen required /mass BOD removed is typically in the range from 0.92 - 1.1 lb O_2/lb BOD (or kg O_2/kg BOD). Higher SRT results in a higher value of mass O_2 required/mass BOD removed.

5. The oxidation of ammonia nitrogen typically requires 4.1 to 4.6 lb of oxygen/lb ammonia Nitrogen oxidized (or kg oxygen/kg ammonia nitrogen oxidized).

Example #8: Calculate the oxygen requirement in lb/hr, the air requirement in SCFM, and the required blower outlet pressure for the wastewater flow used in the previous examples (3.5 MGD wastewater flow with primary effluent BOD of 175 mg/L, primary effluent TKN of 35 mg/L, target effluent BOD of 20 mg/L, and target effluent NH_3-N of 7 mg/L).

Solution: The **Figure 12** and **Figure 13** screenshots on the next page show an Excel spreadsheet with the solution to this example. **Figure 12** shows the wastewater characteristics user inputs needed for the calculation and **Figure 13** shows the diffuser parameters/characteristics inputs needed and the calculation of the oxygen requirement, air requirement, and blower outlet pressure needed for BOD removal and for BOD removal and nitrification.

The user inputs shown in **Figure 12** are those specified in the problem, wastewater flow rate, $Q_o = 3.5$ MGD; primary effluent BOD, $S_o = 175$ mg/L; primary effluent TKN, $TKN_o = 35$ mg/L; target effluent BOD, $S_e = 20$ mg/L and target effluent NH_3-N, $N_e = 7$ mg/L.

Activated Sludge Waste Water Treatment Calculations - S.I. units

4. Oxygen Requirement/Blower Sizing Calculations

Instructions: Enter values in blue boxes. Spreadsheet calculates values in yellow boxes

Values Transferred from Previous Worksheets:

Design ww Flow Rate, Q_o = 5678 m³/d   Prim. Effl. BOD, S_o = 160 g/m³

Additional User Inputs:

Influent TKN, TKN_o = 35 g/m³   Target Effl NH_4-N conc. N_e = 7 g/m³
(needed only if nitrification is to be achieved)   (needed only if nitrification is to be achieved)

Target Effluent BOD, S_e = 20 g/m³

Figure 12. Screenshot of Spreadsheet Solution to Example #8 – Part 1

Source of spreadsheet for screenshot:
www.EngineeringExcelTemplates.com

A. Simplified Estimates Using "Rules of Thumb" - See Notes at Right

1. Inputs: (Values of "Rule of Thumb" Constants to be used in Calculations - See notes at right)

Parameter	Value	Units	Parameter	Value	Units
O_2 needed per lb BOD =	1.20	lb O_2/lb BOD	Depth of Diffusers =	12.0	ft
O_2 needed per lb NH_3-N =	4.57	lb O_2/lb NH_3-N	Standard Temperature =	68	°F
SOTE as Function of Depth =	2.00%	% per ft depth	Standard Pressure =	14.7	psi
AOTE/SOTE =	0.33		Atmospheric Pressure =	14.7	psi
Press. Drop across Diffuser = (from mfr/vendor)	12.0	in W.C.	Air Density at STP =	0.075	lbm/SCF
			O_2 Content in Air =	0.017	lbm/SCF

2. Calculations for BOD Removal Only:

BOD Removal Rate =	188.5	lb/hr	AOTE =	7.9%	
Oxygen Requirement =	226.2	lb/hr	Air Requirement =	2751.8	SCFM
SOTE =	24.0%		Blower Outlet Pressure =	20.3	psia

3. Calculations for BOD Removal and Nitrification:

NH_3-N Removal Rate =	34.1	lb/hr	AOTE =	7.9%	
Oxygen Requirement =	381.9	lb/hr	Air Requirement =	4644.9	SCFM
SOTE =	24.0%		Blower Outlet Pressure =	20.3	psia

Figure 13. Screenshot of Spreadsheet Solution to Example #8 – Part 2

In **Figure 13** oxygen and air requirements are calculated for BOD removal only and for BOD removal and nitrification. Operating conditions will determine whether nitrification takes place or doesn't take place. Both BOD removal and nitrification are carried out by aerobic microorganisms, but it is a different set of microorganisms for each. The nitrifying microorganisms are favored by higher temperatures and higher oxygen concentration in the aeration tank. Thus, at low temperatures and/or relatively low oxygen concentrations, BOD removal will still take place but little nitrification will occur. The optimum D.O. level for nitrification is 3.0 mg/L, significant nitrification still occurs for D.O. levels between 2.0 and 3.0 mg/L, and nitrification ceases for D.O. below 0.5 mg/L. The optimum temperature range for nitrification is between 28°C and 32°C, at 16°C, the nitrification rate is about half of that at 30°C. Below 10°C, the nitrification rate is decreased significantly.

The equations for making the oxygen/air/blower calculations shown in **Figure 13** are as follow:

BOD Removal Rate = (Qo*(So- Se)   NH$_3$-N Removal Rate = Qo*(TKNo - Ne)

O$_2$ Requirement = (BOD Rem. Rate)(lb O$_2$/lb BOD) = (NH$_3$-N Rem. Rate)(lb O2/lb NH$_3$-N)

SOTE = (SOTE %/ft depth)(Diffuser Depth)   AOTE = SOTE(AOTE/SOTE)

Air Requirement = (O$_2$ requirement/AOTE)/(O$_2$ Content in Air)

Blower Outlet Pressure = P$_{atm}$ + Press. Drop across Diffuser + γ_{air}(Diffuser Depth)

The calculated results for **Example #8**, as shown in **Figure 13** are as follows:

For operating conditions in which only BOD removal will take place:

Oxygen Requirement = **226.2 lb/hr**   Air Requirement = **2751.8 SCFM**

Blower Outlet Pressure = **20.3 psi**

For operating conditions in which only BOD removal and nitrification will take place:

Oxygen Requirement = **381.9 lb/hr**   Air Requirement = **4644.9 SCFM**

Blower Outlet Pressure = **20.3 psi**

Alkalinity Requirements also need to be calculated if nitrification is taking place, because alkalinity is used in the nitrification reactions.

Description of the MBBR (Moving Bed Biofilm Reactor)

Initial Development of the Process

The MBBR process for wastewater treatment was invented and initially developed by Professor Hallvard Ødegaard in the late 1980s at the Norwegian University of Science and Technology. Use of this wastewater treatment process has spread rapidly. Per Ødegaard, 2014 (Reference #4 at the end of this book), there were already more than 800 MBBR wastewater treatment plants in more than 50 countries at that time (2014), with about half treating domestic wastewater and about half treating industrial wastewater.

General Description of the Process

The MBBR process is an attached growth biological wastewater treatment process. That is, the microorganisms that carry out the treatment are attached to a solid medium, as in trickling filter or RBC systems. By contrast, in a suspended growth biological wastewater treatment process, like the activated sludge process, the microorganisms that carry out the treatment are kept suspended in the mixed liquor in the aeration tank.

In the conventional attached growth biological treatment processes, like trickling filter or RBC systems the microorganisms are attached to a medium that is fixed in place and the wastewater being treated flows past the surfaces of the medium with their attached biological growth. In contrast, an MBBR process utilizes small plastic carrier media, which are described in more detail in the next section. The MBBR treatment processes typically take place in a tank similar to an activated sludge aeration tank. The carrier media are kept suspended by a diffused air aeration system for an aerobic process or by a mechanical mixing system for an anoxic or anaerobic process, as illustrated in **figure 14** and **figure**

15 below. A sieve is typically used at the tank exit to keep the carrier media in the tank.

Primary clarification is typically used ahead of the MBBR tank. Secondary clarification is also typically used, but there is no recycle activated sludge sent back into the process.

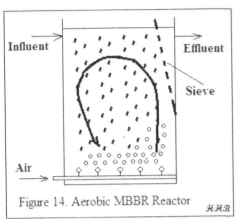

Figure 14. Aerobic MBBR Reactor

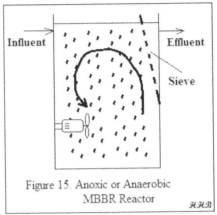

Figure 15. Anoxic or Anaerobic MBBR Reactor

The MBBR Media Support Carrier System

MBBR processes use plastic media support carriers similar to those shown in Figure 16. As shown in **Figure 16**, the carrier is typically designed to have a high surface area per unit volume, so that there is a lot of surface area on which the microorganisms attach and grow. Media support carriers like those shown in Figure 16 are available from numerous vendors. Two properties of the carrier are needed for the process design calculations to be described and discussed in this course. Those properties are the specific surface area in m^2/m^3 and the void ratio. The specific surface area of MBBR carriers is typically in the range from 350 to 1200 m^2/m^3 and the void ratio typically ranges from 60% to 90%. Design values for these carrier properties should be obtained from the carrier manufacturer or vendor.

Figure 16. Typical MBBR Media Support Carriers

MBBR Wastewater Treatment Process Alternatives

The MBBR wastewater treatment process is quite flexible and can be used in several different ways. Figure 17 below shows flow diagrams for the following six alternatives. Note that, as previously mentioned, primary clarification and secondary clarification are shown for all of the process alternatives, but there is no sludge recycle as there is in a conventional activated sludge process.

1. Single stage BOD removal
2. Two stage BOD removal
3. Two stage BOD removal and Nitrification
4. Single stage tertiary Nitrification
5. Pre-Anoxic Denitrification
6. Post-Anoxic Denitrification

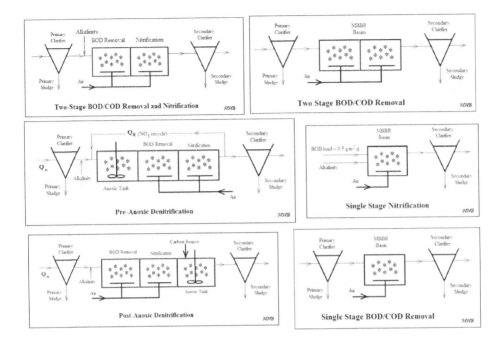

Figure 17. MBBR Wastewater Treatment Process Alternatives

Overview of MBBR Process Design Calculations

The key empirical design parameter used to determine the required MBBR tank size is the surface area loading rate (SALR) in $g/m^2/d$. The g/d in the SALR units refers to the g/d of the parameter being removed and the m^2 in the SALR units refers to the surface area of the carrier. Thus, for BOD removal the SALR would be g BOD/day entering the MBBR tank per m^2 of carrier surface area. For a nitrification reactor, the SALR would be g NH_3-N/day entering the MBBR tank per m^2 of carrier surface area. Finally, for denitrification design, the SALR would be g NO_3-N/day per m^2 of carrier surface area.

For any of these processes, a design value for SALR can be used together with design values of wastewater flow rate and BOD, ammonia or nitrate concentration, to calculate the required carrier surface area in the MBBR tank. The design carrier volume can then be calculated using a known value for the carrier specific surface area (m^2/m^3). Finally, a design value for the carrier fill % can be used to calculate the required tank volume.

Process design calculations for each of the process alternatives shown in **Figure 17** will be covered in the next several sections.

Single Stage BOD Removal MBBR Process Design Calculations

An MBBR single stage BOD removal process may be used as a free-standing secondary treatment process or as a roughing treatment prior to another secondary treatment process, in some cases to relieve overloading of an existing secondary treatment process. In either case the key design parameter for sizing the MBBR tank is the surface area loading rate (SALR), typically with units of $g/m^2/day$, that is g/day of BOD coming into the MBBR tank per m^2 of carrier surface area. Using design values for wastewater flow rate and BOD concentration entering the MBBR tank, the loading rate in g BOD/day can be calculated. Then dividing BOD loading rate in g/day by the SALR in $g/m^2/day$ gives the required carrier surface area in m^2. The carrier fill %, carrier specific surface area, and carrier % void space can then be used to calculate the required carrier volume, tank volume and the volume of liquid in the reactor. A typical flow diagram for a single stage MBBR process for BOD removal is shown in **Figure 18** below.

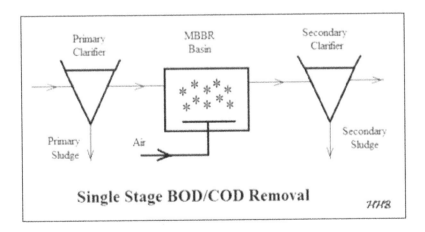

Figure 18. Single Stage MBBR Process for BOD/COD Removal

The equations for making those calculations are as follows:

1. BOD loading rate = $Q * S_o * 8.34 * 453.59$

where: **Q** is the wastewater flow rate into the MBBR reactor in MGD

S_o is the BOD concentration in that influent flow in mg/L
8.34 is the conversion factor from mg/L to lb/MG
453.59 is the conversion factor from lb to g
The calculated **BOD loading rate** will be in g/day.

2. required carrier surf. area = BOD Loading Rate/SALR

where: **BOD Loading Rate** is in g/day

SALR is the design surface area loading rate in $g/m^2/day$
The calculated **required carrier surf. area** will be in m^2.

3. required carrier volume = required carrier surf. area/carrier specific surf. area

where: **required carrier surf. Area** is in m^2
carrier specific surf. Area is in m^2/m^3
The calculated **required carrier volume** will be in m^3.

4. required tank volume = required carrier volume/carrier fill %

where: **required tank volume** will be in the same units as **required carrier volume.**

5. liquid volume in tank
   = required tank volume − [required carrier volume(1 − carrier % void space)]

where: all three volumes will be in the same units.

Note that volumes calculated in m^3 can be converted to ft^3 by multiplying by 3.2808^3 ft^3/m^3.

Although hydraulic retention time (HRT) is not typically used as a primary design parameter for MBBR reactors, it can be calculated at the design wastewater flow rate, if the liquid volume in the tank is known. Also, if a design peak hour factor is specified, then the HRT at peak hourly flow can be calculated as well. The equations for calculating HRT are as follows:

1. **Ave. $HRT_{des\ ave}$ = liquid vol. in tank*7.48/[Q*10^6/(24*60)]**

where: **liquid vol. in tank** is in ft^3
  Q is in MGD
  7.48 is the conversion factor for ft^3 to gal
  10^6 is the conversion factor for MG to gal
  24*60 is the conversion factor for days to min
  Ave. $HRT_{des\ ave}$ will be in min

2. **Ave. $HRT_{peak\ hr}$ = Ave. $HRT_{des\ ave}$/Peak Hour Factor**

where: **Ave. $HRT_{peak\ hr}$** will also be in min

Table 3 below shows typical SALR design values for BOD removal in MBBR reactors. Reference #3 at the end of this book (Odegaard, 1999) is the source for the values in **Table 3**.

Table 3. Typical Design SALR Values for BOD Removal

Typical Design Values for MBBR reactors at 15°C		
Purpose	Treatment Target % Removal	Design SALR g/m^2-d
BOD Removal		
High Rate	75 - 80 (BOD_7)	25 (BOD_7)
Normal Rate	85 - 90 (BOD_7)	15 (BOD_7)
Low Rate	90 - 95 (BOD_7)	7.5 (BOD_7)

Example #10: A design wastewater flow of 1.5 MGD containing 175 mg/L BOD (in the primary effluent) is to be treated in an MBBR reactor.

a) What is the BOD loading rate to the reactor in g/day?
b) What would be a suitable design SALR to use for a target % removal of 90-95% ?
c) If the MBBR carrier has a specific surface area of 600 m^2/m^3 and design carrier fill % of 40%, what would be the required volume of carrier and required MBBR tank volume?
d) If the design carrier % void space is 60%, what would be the volume of liquid in the MBBR reactor.
e) If the design peak hour factor is 4, calculate the average hydraulic retention time at design average wastewater flow and at design peak hourly wastewater flow.

Solution:

a) The BOD loading rate will be (1.5 MGD)(175 mg/L)(8.34 lb/MG/mg/L) = 2189 lb/day = (2189 lb/day)*(453.59 g/lb) = **993,000 g BOD/day**

b) From Table 1 above, a suitable design SALR value for BOD removal with a target BOD removal of 90–95% would be **7.5 g/m^2/day**

c) Required carrier surface area = (993,000 g/day)/(7.5 g/m^2/day) = 132,403 m^2.
Required carrier volume = 308,940 m^2/600 m^2/m^3. = 220.7 m^3

For 40% carrier fill: Required tank volume = 220.7 m^3/0.40 = **551.7 m^3**.

d) The volume of liquid in the reactor can be calculated as:
tank volume – [carrier volume(1 – void %)], Thus the volume of liquid is: 551.7 – [220.7(1 – 0.60)] = **463.4 m^3** = 463.3*(3.2808^3) = **16,365 ft^3**

e) The HRT at design ave ww flow can be calculated as:

$HRT_{des\ ave}$ = reactor liquid volume*7.48/[Q*10⁶/(24*60)]
= 16,365*7.48/[1.5*10⁶/(24*60)] = **118 min**

$HRT_{peak\ hr}$ = $HRT_{des\ ave}$/peak hour factor = 118/4 = **29 min**

Use of an Excel Spreadsheet

The type of calculations just discussed can be done conveniently using an Excel spreadsheet. The screenshot in **Figure 19** below shows an example Excel spreadsheet set up to make the calculations just described for **Example #10**.

The worksheet shown in **Figure 19** is set up for user input values to be entered in the blue cells with the values in the yellow cells calculated by the spreadsheet using the equations presented and discussed above. The values calculated by the spreadsheet for the following parameters are the same as that shown in the solution to **Example #10** above: i) the BOD loading in g/day, ii) the required tank volume, iii) the liquid volume in the tank, iv) the hydraulic retention time at design average flow, and v) the hydraulic retention time at peak hourly flow.

You may have noticed that there are a few additional calculations in the worksheet screenshot. If the user selected tank shape as "rectangular," the worksheet calculates the tank length and width for the calculated required tank volume. If the user selected tank shape is cylindrical, the worksheet will calculate the tank diameter. These calculations simply use the formulas for the volume of a rectangular tank (V = L*W*H) or for the volume of a cylindrical tank (V = πD²H/4), with user entered values for L:W ratio being used for rectangular tank calculations and the user entered value for the liquid depth in the tank, H, being used by both. The worksheet also calculates an estimated effluent BOD concentration. The effluent BOD calculation is discussed in the next section.

MBBR Process Design Calculations - U.S. units
Single-Stage Process for BOD Removal

Instructions: Enter values in blue boxes. Spreadsheet calculates values in yellow boxes

1. General Inputs

			Data points for SARR/SALR vs SALR		
Design ww Flow Rate, Q =	1.5	MGD	SALR ($g/m^2/d$):	7.5	15.0
Prim. Effl. BOD, S_o =	175	mg/L	SARR/SALR:	0.925	0.875
Peak Hour Factor =	4		(default values above are based on the table of typical values of % BOD removal vs SALR at the right)		

Design Value of BOD Surface			Slope, SARR/SALR vs SALR:	-0.007
Area Loading Rate (SALR) =	7.5	$g/m^2/d$	Intercept, SARR/SALR vs SALR:	0.975
See information on typical design			Est. of SARR/SALR Rato =	0.925
values for SALR below right.			(Surf. Area Removal Rate/Surf. Area Loading Rate) (for SALR value specified at left)	

2. Calculation of Carrier Volume and Required Tank Volume & Dimensions

Inputs

			Liquid Depth in Tank =	8	ft
Carrier Spec. Surf. Area =	600	m^2/m^3	Tank L:W ratio =	1.5	
(value from carrier mfr/vendor)			(target L:W - only used if tank is rectangular)		
Design Carrier Fill % =	40%		Click on green box and then on arrow to Select Tank Shape:	rectangular	
(Carrier fill % is typically between 30% and 70%. Lower values are more conservative, allowing future capacity expansion or reduction of SALR by adding more carrier.)			Carrier % Void Space =	60%	
			(from carrier mfr/vendor - only needed to calculate hydraulic detention time)		

Calculations

BOD Daily Loading =	2189	lb/day	Calculated Tank Volume =	551.7	m^3
	993,022	g/day	=	19,482	ft^3
			=	145,723	gal
Carrier Surf. Area needed =	132,403	m^2			
Calculated Carrier Volume =	220.7	m^3	Calculated Tank Width =	40.3	ft
			Calculated Tank Length =	60.4	ft
Tank Liquid Volume =	16,365	ft^3	Nominal Hydraulic Retention Time at		
	463.4102	16364.58	Design Average Flow =	118	min
Estimate of BOD Surface Area			Peak Hourly Flow =	29	min
Removal Rate, SARR =	6.94	$g/m^2/d$			
Est. of BOD Removal Rate	918,545	g/day	Calculated Effl BOD Conc.:	13	mg/L
	2025.1	lb/day	If the calculated Effl. BOD conc. is too high, the design value of SALR (in cell C13) should be reduced.		

Figure 19. Screenshot of MBBR Process Design Calculations – Single Stage BOD

Estimation of Effluent Concentration

Use of an estimated surface area removal rate (SARR) allows calculation of the estimated effluent concentration of the parameter being removed. That is, for BOD removal, the estimated effluent BOD concentration can be calculated. For nitrification, the estimated effluent ammonia nitrogen concentration can be calculated and for denitrification, the estimated effluent nitrate nitrogen concentration can be calculated.

Based on graphs and tables provided in several of the references at the end of this book, the SARR/SALR ratio for all of the different types of MBBR treatment being covered in this course ranges from about 0.8 to nearly 1.0 over the range of SALR values typically used. The SARR/SALR ratio is nearly one at very low SALR values and decreases as the SALR value increases.

The upper right portion of the screenshot on the previous page illustrates an approach for estimating a value for the SARR/SALR ratio for a specified design value of SALR. In the four blue cells at the upper right, two sets of values for SARR/SALR and SALR are entered. In this case they are based on the typical values of % BOD removal vs SALR in Table 1 above. In the yellow cells below those entries, the slope and intercept of a SARR/SALR vs SALR straight line are calculated using the Excel SLOPE and INTERCEPT functions. Then the SARR/SALR ratio is calculated for the specified design value of SALR.

Note that the ratio SARR/SALR is equal to the % BOD removal expressed as a fraction. This can be shown as follows:

BOD removal rate in g/day = (SARR in $g/m^2/d$)(Carrier Surf. Area in m^2)

BOD rate into plant in g/day = (SALR in $g/m^2/d$)(Carrier Surf. Area in m^2)

% BOD removal = (BOD removal rate/BOD rate into plant)*100%

= (100%)(SARR* Carrier Surf Area)/(SALR*Carrier Surf Area)

= **(SARR/SALR)100%**

Thus the value of **0.925** for the **SARR/SALR** ratio at **SALR = 7.5 $g/m^2/d$** was obtained from **Table 3** above as the midpoint of the 90-95% estimated

% BOD removal for **SALR = 7.5 g/m²/d**. Similarly, the value of **0.875** for the **SARR/SALR** ratio at **SALR = 15 g/m²/d** was obtained from **Table 3** above as the midpoint of the 85-90% estimated % BOD removal for **SALR = 15 g/m²/d**.

At the bottom of the screenshot worksheet, the estimated value of the surface area removal rate (SARR) is calculated. It is used to calculate an estimated BOD removal rate in g BOD/day and lb BOD/day. Then an estimate of the effluent BOD concentration is calculated. The equations for these calculations are as follows:

1. estimated SARR = (calculated SARR/SALR)(design value of SALR)

2. estimated BOD removal rate = (estimated SARR)(carrier surface area)

3. estimated effluent BOD conc.
   = [(BOD loading rate - estimated BOD removal rate)/Q_o]/8.34

Example #11: Calculate the estimated effluent BOD concentration for the wastewater flow described in **Example #10** being treated in the MBBR reactor sized in **Example #10**.

Solution: The solution is included in the **Figure 19** spreadsheet screenshot that was used for the solution to **Example #10** above. The pair of points for **SARR/SALR vs SALR** that were discussed above and are shown on the **Figure 19** screenshot lead to the following values for the slope and intercept for the **SARR/SALR vs SALR** line: **Slope = - 0.007,   Intercept = 0.975**.

Thus the estimated **SARR/SALR** ratio for the given **SALR** value of **7.5 g/m²/d** would be calculated as: **SARR/SALR = - (0.007)(7.5) + 0.975 = 0.925**

The SARR value can be calculated as:

SARR = (SARR/SALR)(SALR) = (0.925)(7.5) = 6.94 g/m²/d

Then, the estimated BOD removal rate can be calculated as:

est BOD removal rate = (est SARR)(carrier surface area)

= (6.94 g/m²/d)(132,403 m²) = 918,545 g/d = 918,545/453.59 lb/day

est BOD removal rate = 2025.1 lb/day

The estimated effluent BOD concentration can then be calculated from the equation:

est effluent BOD conc. = [(BOD loading rate - est BOD removal rate)/Q_o]/8.34

Substituting calculated and given values:

Est. effluent BOD conc. = [(2189 – 2026.1)/1.5]/8.34 = **13 mg/L**

Note that this **13 mg/L** value for the **estimated effluent BOD** concentration is shown near the bottom of the **Figure 19** spreadsheet screenshot.

Two-Stage BOD Removal MBBR Process Design Calculations

A two stage MBBR BOD removal process may be used instead of a single stage process. In this case, a high SALR "roughing" treatment will typically be used for the first stage and a lower SALR will typically be used for the second stage. This will result in less total tank volume needed for a two-stage process than for a single stage process. Also, a two-stage MBBR process can typically achieve a lower effluent BOD concentration than a single stage MBBR process. A typical flow diagram for a two-stage MBBR process for BOD removal is shown in **Figure 20** below

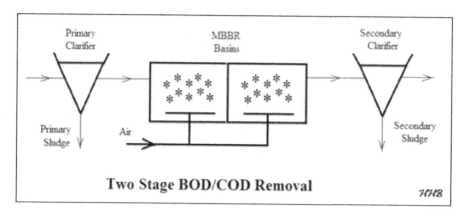

Figure 20. Two Stage MBBR Process for BOD/COD Removal

The process design calculations for a two stage MBBR process are essentially the same for each of the stages as for the single stage process, as described in the previous section. These calculations are illustrated in **Example #12**.

Example #12: A design wastewater flow of 1.5 MGD containing 175 mg/L BOD (in the primary effluent) is to be treated for BOD removal in a two-stage MBBR reactor. The SALR for the first stage is to be 25 g/m²/d and the design SALR for the second stage is to be 7.5 g/m²/d.

a) For the first stage calculate each of the following:
   i) The BOD loading
   ii) The required carrier volume for a carrier with specific surface area of 600 m²/m³
   iii) The required MBBR tank volume for a design carrier fill % of 40%
   iv) The volume of liquid in the MBBR reactor for design carrier % void space of 60%.
   v) The average hydraulic retention time at design average wastewater flow and at design peak hourly flow if the design peak hour factor is 4.
   vi) The estimated effluent BOD concentration from the first stage.

b) Calculate the same parameters for the second stage.

Solution: The solution as calculated with an Excel spreadsheet is shown in the spreadsheet screenshot in **Figure 21** and **Figure 22**. **Figure 21**, which is the top part of the spreadsheet, shows primarily the user input values. It also includes the calculation of the slope and intercept for the SARR/SALR vs SALR equation and the calculation of the estimated SARR for each stage. These calculations and the resulting SARR/SALR values are the same as those discussed above for the single-stage BOD removal MBBR process. The resulting values for SARR/SALR are **0.775** for the first stage with SALR = 25, and **0.925** for the second stage with SALR = 7.5.

Figure 22 is the bottom part of the spreadsheet and shows the calculated values as follows.

   a) For the first stage:

   i) The BOD loading rate will be (1.5 MGD)(175 mg/L)(8.34 lb/MG/mg/L) = 2189 lb/day = (2189 lb/day)*(453.59 g/lb) = **993,022 g BOD/day**

   ii) Required carrier surface area = (993,022 g/day)/(25 g/m²/day) = 39,721 m².

   Required carrier volume = 39,721 m²/600 m²/m³. = 66.20 m³

iii) For 40% carrier fill: Required tank volume = 66.2 m³/0.40 = **165,5 m³**.

iv) The volume of liquid in the reactor can be calculated as: tank volume – [carrier volume(1 – void %)], Thus the volume of liquid is: 165.5 – [66.20(1 – 0.60)] = **139.02 m³**. = 139.02(3.2808³) = **4910 ft³**

v) The HRT at design ave ww flow can be calculated as:
$HRT_{des\ ave}$ = reactor liquid volume*7.48/[Q*10⁶/(24*60)] = 4910*7.48/[1.5*10⁶/(24*60)] = **35 min**

$HRT_{peak\ hr}$ = $HRT_{des\ ave}$/peak hour factor = 35/4 = **9 min**

vi) Calculation of the estimated effluent BOD concentration from the first stage as shown above for the single stage process gives a value of **39 mg/L**.

b) For the second stage:

i) The NH_3-N loading rate will be (1.5 MGD)(39 mg/L)(8.34 lb/MG/mg/L) = 492.6 lb/day = (492.6 lb/day)*(453.59 g/lb) = **223,430 g BOD/day**

ii) Required carrier surface area = (223,430 g/day)/(7.5 g/m²/day) = 29,791 m².

Required carrier volume = 29,791 m²/600 m²/m³. = 49.65 m³

iii) For 40% carrier fill: Required tank volume = 288.55 ft³/0.40 = **124.1 m³**.

iv) The volume of liquid in the reactor can be calculated as: tank volume – [carrier volume(1 – void %)], Thus the volume of liquid is: 124.1 – [49.65(1 – 0.60)] = **104.3 m³** = 104.3(3.2808³) = **3682 ft³**

MBBR Process Design Calculations - U.S. units
Two-Stage Process for BOD Removal

Instructions: *Enter values in blue boxes. Spreadsheet calculates values in yellow boxes*

I. Wastewater Parameter Inputs

1. Parameters for Both First and Second Stage

Design ww Flow Rate, Q = 1.5 MGD   Peak Hour Factor = 4

2. Parameters for First Stage:

Data points for SARR/SALR vs SALR

SALR (g/m²/d):	7.5	25.0
SARR/SALR:	0.925	0.78

Prim. Effl. BOD, S_{o1} = 175 mg/L

(default values above are based on the table of typical values of % BOD removal vs SALR at the right)

Design Value of BOD Surface
Area Loading Rate (SALR) = 25 g/m²/d

See information on typical design values for SALR below right.

Slope, SARR/SALR vs SALR: -0.009
Intercept, SARR/SALR vs SALR: 0.989
Est. of SARR/SALR Rato = 0.775
(Surf. Area Removal Rate/Surf. Area Loading Rate)

3. Parameters for Second Stage:

Design Value of BOD Surface
Area Loading Rate (SALR) = 7.5 g/m²/d

See information on typical design values for SALR to the right.

Est. of SARR/SALR Rato = 0.925
(Surf. Area Removal Rate/Surf. Area Loading Rate)

II. Carrier Parameter and Tank Shape Inputs for both First and Second Stages

Carrier Spec. Surf. Area = 600 m²/m³
(value from carrier mfr/vendor)

Click on green box and then on arrow to Select Tank Shape: **rectangular**

Liquid Depth in Tank = 8 ft
Tank L:W ratio = 1.5
(target L:W - only used if tank is rectangular)

Carrier % Void Space = 60%
(from carrier mfr/vendor - only needed to calculate hydraulic detention time)

Figure 21. Screenshot of MBBR Process Design Calculations Two Stage BOD Removal – Part 1

III. Calculation of Carrier Volume and Required Tank Volume & Dimensions

1. First Stage Calculations (BOD Removal)

(Carrier fill % is typically between 30% and 70%. Lower values are more conservative, allowing future capacity expansion or reduction of SALR by adding more carrier.)

Design Carrier Fill % =	40%	(for first stage)	
BOD Daily Loading =	2189	lb/day	
	993,022	g/day	Calculated Tank Volume = 165.5 m³
Carrier Surf. Area needed =	39,721	m²	= 5844.7 ft³
Calculated Carrier Volume =	66.20	m³	= 43718 gal
Tank Liquid Volume =	4910	ft³	Calculated Tank Width = 22.1 ft
	139.0231	4909.374	Calculated Tank Length = 33.1 ft
Estimate of BOD Surface Area			Nominal Hydraulic Retention Time at
Removal Rate, SARR =	19.38	g/m²/d	Design Average Flow = 35 min
Est. of BOD Removal Rate:	769,592	g/day	Peak Hourly Flow = 9 min
	1696.7	lb/day	Calculated Effl BOD Conc.: 39 mg/L (from First Stage)

2. Second Stage Calculations (BOD Removal)

Design Carrier Fill % =	40%	(for second stage)	
BOD Daily Loading =	492.6	lb/day	Calculated Tank Volume = 124.1 m³
	223,430	g/day	= 4383.5 ft³
Carrier Surf. Area needed =	29,791	m²	= 32789 gal
Calculated Carrier Volume =	49.65	m³	Calculated Tank Width = 19.1 ft
Tank Liquid Volume =	3682	ft³	Calculated Tank Length = 28.7 ft
	104.2673003	3682.031	Nominal Hydraulic Retention Time at
Estimate of BOD Surface Area			Design Average Flow = 26 min
Removal Rate, SARR =	6.94	g/m²/d	Peak Hourly Flow = 7 min
Est. of BOD Removal Rate:	206,673	g/day	Calculated Effl BOD Conc.: 3.0 mg/L (from Second Stage)
	455.6	lb/day	
1st stage tank volume -			If the calculated Effl. BOD conc. is too high for either
- 2nd stage tank volume =	41.4		stage, the design value of SALR should be reduced for that stage.

To make the 2nd stage tank volume the same as the first stage tank volume, use Excel's Goal Seek process to set cell C66 equal to zero by changing the value in cell C54.

Figure 22. Screenshot of MBBR Process Design Calculations Two Stage BOD Removal – Part 2

v) The HRT at design ave ww flow can be calculated as:
$HRT_{des\ ave}$ = reactor liquid volume*7.48/[Q*10⁶/(24*60)] =
3682*7.48/[1.5*10⁶/(24*60)] = **26 min**

$HRT_{peak\ hr}$ = $HRT_{des\ ave}$/peak hour factor = 26/4 = **7 min**

Calculation of the estimated effluent BOD concentration from the second stage as shown above for the single stage process gives a value of **3.0 mg/L**.

Example #13: Compare the MBBR tank volume, carrier surface area, and estimated effluent BOD concentration for the single stage BOD removal process in **Example #10** and the same WW flow and BOD with a two-stage BOD removal process as calculated in **Example #12**.

Solution: The results are summarized below:

	Single Stage Process	Two-Stage Process
MBBR Volume:	19,482 ft^3	10,228 ft^3
Carrier Surf. Area:	132,403 m^2	69,512 m^2
Est. Effl. BOD:	13 mg/L	3 mg/L

Note that the two-stage process requires only about half of the tank volume and half of the carrier quantity in comparison with the single stage process, while achieving a significantly lower estimated effluent BOD.

Single Stage Nitrification MBBR Process Design Calculations

An MBBR single stage nitrification process would typically be used as a tertiary treatment process following some type of secondary treatment that reduced the BOD to a suitable level. A typical flow diagram for a single stage MBBR process for nitrification is shown in **Figure 23** below. As shown on the diagram, the BOD level should be low enough so that the BOD load to the nitrification process is less than 0.5 g/m^2/day. Note that alkalinity is used in the nitrification process and thus alkalinity addition is typically required.

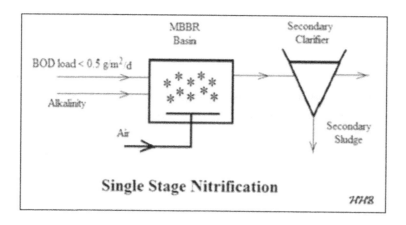

Figure 23. Single Stage MBBR Process for Nitrification

The process design calculations for this single stage MBBR process are similar to those used for the BOD removal processes, but the design SALR value for nitrification can be calculated rather than being selected from a table of typical values, as was done for BOD removal. The design SALR can be calculated using a kinetic model for the surface area removal rate (SARR) as a function of the dissolved oxygen concentration in the MBBR reactor and the bulk liquid ammonia nitrogen concentration, which is equal to the effluent ammonia nitrogen concentration assuming completely mixed conditions in the MBBR tank.

The kinetic model to be discussed here is from Metcalf and Eddy (2014), Figure 9-25 [attributed to Odegaard (2006)] and Equation 9-48. This figure and equation will now be shown and discussed briefly. **Figure 24** below was prepared based on Metcalf and Eddy's Figure 9-25 and their Equation 9-48 (shown below). Note that the figure and equation are for operation at 15°C. Correction of the SARR and SALR for some other operational temperature can be done with the equation: $SARR_T = SARR_{15}\theta^{(T-15)}$ where T is the design operational temperature in °C. From Salvetti, et.al (2006): $\theta = \mathbf{1.058}$ for D.O. limited conditions and $\theta = \mathbf{1.098}$ if ammonia nitrogen concentration is the limiting factor.

Equation 9-48 from Metcalf and Eddy (2014) is:

SARR = [$N_e/(2.2 + N_e)$]*3.3 , N_e = effluent ammonia N conc.

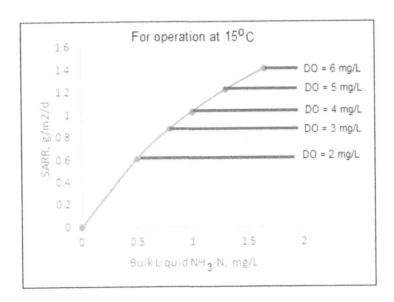

Figure 24. Adapted from Metcalf & Eddy (2014), Fig 9-25

The horizontal lines in **Figure 24** show the nitrification **SARR** under DO limiting conditions for each of the DO levels shown. The **SARR** will be DO limited when the **NH3-N** concentration is above the value at the left end of the horizontal line for each D.O. level. When the **NH3-N** concentration is below that value, then the **SARR** is ammonia concentration limited and the **SARR** is a function of the effluent **NH$_3$-N** concentration per the equation: **SARR = [N$_e$/(2.2 + N$_e$)]*3.3.**

As shown in **Chapter 7** for BOD removal, **SALR/SARR = % BOD removal**. Similarly, for nitrification, **SALR/SARR = % NH$_3$-N removal**. After the **SARR** has been determined, the **SALR** can be calculated as: **SALR = SARR/% removal.**

The maximum **SARR** for each of the D.O. levels shown in **Figure 24** are shown in **Table 4** below, along with the ammonia nitrogen concentration above which the **SARR** will be at that maximum value.

Table 4. Values of **SARR$_{max}$** and **NH$_3$-N$_e$ @ SARR$_{max}$**

D.O.	SARR$_{max}$	min NH$_3$-N$_e$ @ SARR$_{max}$
mg/L	g/m^2/d	mg/L
2	0.61	0.5
3	0.88	0.8
4	1.03	1
5	1.23	1.3
6	1.41	1.65

Example #14: A design flow of 0.2 MGD has the following characteristics: 15 mg/L BOD, 25 mg/L NH$_3$-N, and alkalinity of 140 mg/L as CaCO$_3$. This flow is to be treated in a single stage nitrification MBBR reactor. The target effluent NH$_3$-N is 3.3 mg/L. The dissolved oxygen is to be maintained at 3.0 mg/L in the MBBR reactor. The design minimum wastewater temperature is to be 45°F.

a) Determine an appropriate NH_3-N surface area loading rate (SALR) for this nitrification process, in g NH_3-N/m²/d.
b) What is the NH_3-N loading rate to the reactor in g/day?
c) If the MBBR carrier has a specific surface area of 600 m²/m³ and design carrier fill % of 40%, what would be the required volume of carrier and required MBBR tank volume?
d) Calculate the BOD surface area loading rate (SALR) to ensure that it is less than 0.5 g BOD/m²/day.
e) If the design carrier % void space is 60%, what would be the volume of liquid in the MBBR reactor.
f) If the design peak hour factor is 4, calculate the average hydraulic retention time at design average wastewater flow and at design peak hourly wastewater flow.
g) The alkalinity requirement in lb/day as $CaCO_3$ and in lb/day $NaHCO_3$.

Solution:

The solution is shown in the spreadsheet screenshots in **Figure 25** and **Figure 26** below. A summary of the calculations is as follows:

a) The D.O. limited **SARR** can be obtained as **SARR**$_{max}$ from **Table 4** for the specified D.O level, and the minimum ammonia nitrogen concentration for that **SARR** value can be obtained from the same table. The values from **Table 4**, for a D.O. of 3.0 are: **SARR**$_{max}$ = **0.88 g/m²/d** and minimum NH_3-N_e for **SARR**$_{max}$ = **0.80 mg/L.** (in the worksheet shown in the **Figure 25** screenshot, these two values are obtained using Excel's VLOOKUP function from a table like **Table 4**, above that is on the worksheet.

The **SARR** for the design D.O. and ammonia nitrogen removal at 15°C will then be equal to **SARR**$_{max}$ if the target effluent ammonia nitrogen concentration is greater than the 0.80 mg/L value determined above. If the target effluent ammonia nitrogen concentration is less than 0.80 mg/L, then the SARR needs to be calculated using Metcalf & Eddy's equation 9-48. In this case, the target effluent NH_3-N of 3.3 mg/L is greater than 0.8 mg/L, so the SARR at 15°C is 0.88 g/m²/d.

The design value for the SARR at the design minimum wastewater temperature can then be calculated as: $SARR_T = SARR_{15}\theta^{(T-15)}$, where the WW temperature must be in °C. Since this case has D.O. limited conditions, $\theta = 1.058$. Carrying out this calculations gives: **design value of SALR = 0.65 g/m²/d**.

b) The ammonia nitrogen loading rate will be (0.2 MGD)(25 mg/L)(8.34 lb/MG/mg/L) = 41.7 lb/day = (41.7 lb/day)*(453.59 g/lb) = **18,915 g BOD/day**

c) Required carrier surface area = (18,915 g/day)/(0.65 g/m²/day) = 28,925 m².

   Required carrier volume = 28,925 m²/600 m²/m³. = 48.209 m³ = (48.209 m³)(3.2808³ ft³/m³) = **1702 ft³**.

   For 40% carrier fill: Required tank volume = 1702 ft³/0.40 = **4256 ft³**.

d) The BOD SALR will be (0.2 MGD)(15 mg/L)(8.34 lb/MG/mg/L)(453.59)/(28925 m²) = **0.39 g/m²/day** (Note that this is less than 0.5 g/m²/day as required.)

e) The volume of liquid in the reactor can be calculated as: tank volume – [carrier volume(1 – void %)], Thus the volume of liquid is: 4256 – [1702(1 – 0.60)] = **3575 ft³**.

f) The HRT at design ave ww flow can be calculated as:
   $HRT_{des\ ave}$ = reactor liquid volume*7.48/[Q*10⁶/(24*60)] = 3575*7.48/[0.2*10⁶/(24*60)] = **193 min**

   $HRT_{peak\ hr}$ = $HRT_{des\ ave}$/peak hour factor = 193/4 = **48 min**

g) Calculation of the alkalinity requirement is shown in the **Figure 26** screenshot and is calculated as follows: Using the equivalent weight of $CaCO_3$ as 50, the equivalent weight of $NaHCO_3$ as 84, the alkalinity use for nitrification as 7.14 g $CaCO_3$/g NH_3-N and the target effluent alkalinity as 80 mg/L as $CaCO_3$, give the calculated alkalinity requirement as 118.5 mg/L as $CaCO_3$. The rate of alkalinity addition needed can then be calculated as: (0.2 MGD)(118.5 mg/L)*8.34 = **197.7 lb/day as $CaCO_3$**. Multiplying this by the ratio of the equivalent weight of $NaHCO_3$ to the equivalent weight of $CaCO_3$ gives the daily $NaHCO_3$ requirement as **332.1 lb/day $NaHCO_3$**.

MBBR Process Design Calculations - U.S. units
Single Stage Tertiary Nitrification Process

Instructions: *Enter values in blue boxes. Spreadsheet calculates values in yellow boxes*

1. General Inputs

			Peak Hour Factor =	4	
			Target Effl NH$_4$-N Conc, **NH$_4$-N$_e$** =	3.3	mg/L
Design ww Flow Rate, **Q** =	0.2	MGD	Min Design Temp., **T** =	45	°F
Influent NH$_4$-N Conc. =	25	mg/L			
Influent BOD, **S$_o$** =	15	mg/L	Click on cell H10 and on arrow to select D.O Con		
Prim. Effl. Alkalinity =	140	mg/L as CaCO$_3$	D.O Conc. in Reactor =	3.0	mg/L

2. Preliminary Calculations - Design SALR value

% NH$_4$-N removal =	87%		NH$_4$-N$_e$ @ SARR$_{max}$ =	0.80	mg/L
Maximum SARR =	0.88	g/m²/d	SARR @ NH$_4$-Ne, 15°C, **SARR$_{15}$** =	0.88	g/m²/d
			SARR @ NH$_4$-Ne, T °C, **SARR$_T$** =	0.57	g/m²/d
SARR Temp. Coeff, **θ** =	1.058				
			Design Value for **SALR** =	0.65	g/m²/d

3. Calculation of Carrier Volume and Required Tank Volume & Dimensions

Inputs

Carrier Spec. Surf. Area =	600	m²/m³	Liquid Depth in Tank =	8	ft
(value from carrier mfr/vendor)			Tank L:W ratio =	1.5	
			(target L:W - only used if tank is rectangular)		
Design Carrier Fill % =	40%		Click on green box and then on		
(Carrier fill % is typically between 30% and			arrow to Select Tank Shape:	rectangular	
70%. Lower values are more conservative,			Carrier % Void Space =	60%	
allowing future capacity expansion or			(from carrier mfr/vendor - only needed to		
reduction of SALR by adding more carrier.)			calculate hydraulic detention time)		

Calculations

			Calculated Tank Volume =	120.5	m³
NH$_3$-N Daily Loading =	41.7	lb/day	=	4256.2	ft³
	18915	g/day	=	31837	gal
Carrier Surf. Area needed =	28925	m²	Calculated Tank Width =	18.8	ft
Calculated Carrier Volume =	48.209	m³	Calculated Tank Length =	28.2	ft
Tank Liquid Volume =	3575.2	ft³	Nominal Hydraulic Retention Time at		
			Design Average Flow =	193	min
BOD Surf. Loading Rate (SALR):	0.39	g/m²/d	Peak Hourly Flow =	48	min
(should be < 0.5 g/m²/d in order to achieve a good nitrification rate)					

Fig. 25. Screenshot of MBBR Process Design Calculations – Single Stage Nitrification – Part 1

5. Alkalinity Requirement

Input: Target Effluent Alkalinity = [80] mg/L

Constants needed for Calculations:

Equiv Wt. of $CaCO_3$ = [50] g/equiv   Equiv Wt. of $NaHCO_3$ = [84] g/equiv

Alkalinity used for Nitrification = [7.14] g $CaCO_3$/g NH_3-N

Calculations

Alkalinity to be added = [118.5] mg/L as $CaCO_3$
Daily Alkalinity Requirement = [197.7] lb/day as $CaCO_3$

For sodium bicarbonate use to add alkalinity:

Daily $NaHCO_3$ Requirement = [332.1] lb/day $NaHCO_3$

Fig. 26. Screenshot of MBBR Process Design Calculations – Single Stage Nitrification – Part 2

Two-Stage BOD Removal and Nitrification Process Design Calculations

A two stage MBBR process may also be used to achieve both BOD removal and nitrification. Nitrification with an MBBR process requires a rather low BOD concentration. Thus the first stage for this process is used for BOD removal and the second stage is used for nitrification. A typical flow diagram for a two stage MBBR process for BOD removal and nitrification is shown in **Figure 27** below. As in the single stage nitrification process alkalinity is used for nitrification, so alkalinity addition is typically required.

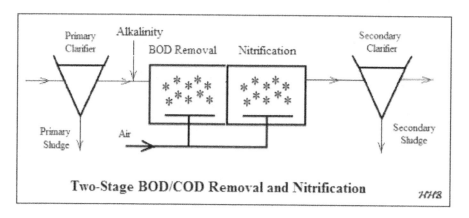

Figure 27. Two Stage MBBR Process for BOD Removal and Nitrification

Table 5 shows typical design values for the SALR (surface area loading rate) for the BOD removal stage and for the nitrification stage. The source for the values in this table is the Odegaard reference #2 below. The design SALR value for nitrification, however, will be calculated based on the design D.O concentration and the target effluent NH_3-N concentration, as it was for the single-stage nitrification process.

Table 5. Typical Design SALR Values for Nitrification

Typical Design Values for MBBR reactors at 15°C		
Purpose	Treatment Target % Removal	Design SALR g/m^2-d
Nitrification BOD removal stage Effl. NH$_3$-N > 3 mg/L Effl. NH$_3$-N < 3 mg/L	90 - 95 (BOD$_7$) 90 (NH$_3$-N) 90 (NH$_3$-N)	6.0 (BOD$_7$) 1.00 (NH$_3$-N) 0.45 (NH$_3$-N)

The process design calculations for this two stage MBBR process are essentially the same as those described above for the previous examples. **Table 3** will be used to obtain values for SARR/SALR vs SALR for the BOD removal stage as in the previous examples. The design SALR for the nitrification stage will be calculated as it was for the single stage nitrification process.

The process design calculations for this two-stage process are illustrated in **Example #15**.

Example #15: A design wastewater flow of 1.5 MGD has the following primary effluent characteristics: 175 mg/L BOD, 35 mg/L TKN, and alkalinity of 140 mg/L as CaCO$_3$. The design minimum wastewater temperature is 45°F. This flow is to be treated for BOD removal and nitrification in a two stage MBBR reactor. The design SALR for the first stage is to be 6 g BOD/m^2/d, the target effluent NH3-N from the second stage is to be 3.3 mg/L and the design DO level in the nitrification stage is to be 3.0 mg/L.

a) For the first (BOD removal) stage calculate each of the following:

i) The BOD loading

ii) The required carrier volume for a carrier with specific surface area of 600 m²/m³
iii) The required MBBR tank volume for a design carrier fill % of 40%
iv) The volume of liquid in the MBBR reactor for design carrier % void space of 60%.
v) The average hydraulic retention time at design average wastewater flow and at design peak hourly flow if the design peak hour factor is 4.
vi) The estimated effluent BOD concentration from the first stage.

b) For the second (Nitrification) stage calculate each of the following:

i) An appropriate NH_3-N surface area loading rate (SALR) to be used for this nitrification process, in g NH_3-N/m²/d.
ii) The nitrate loading
iii) The required carrier volume for a carrier with specific surface area of 600 m²/m³
iv) The required MBBR tank volume for a design carrier fill % of 40%
v) The volume of liquid in the MBBR reactor for design carrier % void space of 60%.
vi) The average hydraulic retention time at design average wastewater flow and at design peak hourly flow if the design peak hour factor is 4.
vii) The alkalinity requirement in lb/day as $CaCO_3$ and in lb/day $NaHCO_3$.

Solution: The solution as calculated with an Excel spreadsheet is shown in the spreadsheet screenshot in **Figure 28** and **Figure 29**. **Figure 28**, which is the top part of the spreadsheet, shows primarily the user input values. It includes the calculation of the slope and intercept for the SARR/SALR vs SALR equation and the calculation of the estimated SARR for the BOD removal stage. It also includes calculation of the design SALR value for the nitrification stage. These calculations are carried out as discussed above for the single-stage BOD removal and nitrification MBBR processes. The resulting values for SARR/SALR are **0.935** for the first stage with SALR = 6 g BOD/m²/d, and **0.63** g/m2/d for the second stage SALR.

Figure 29 is the bottom part of the spreadsheet and shows the calculated values as follows.

a) For the first stage:

i) The BOD loading rate will be (1.5 MGD)(175 mg/L)(8.34 lb/MG/mg/L) = 2,189 lb/day = (300.2 lb/day)*(453.59 g/lb) = **993,022 g BOD/day**

ii) Required carrier surface area = (993022 g/day)/(6 g/m^2/day) = 165,504 m^2.

Required carrier volume = 165,504 m^2/600 m^2/m^3. = 275.8 m^3

iii) For 40% carrier fill: Required tank volume = 275.8 m^3/0.40 = **689.6 m^3**.

iv) The volume of liquid in the reactor can be calculated as:
tank volume − [carrier volume(1 − void %)], Thus the volume of liquid is: 689.6 − [275.8(1 − 0.60)] = **579.3 m^3** = 579.3(3.2808^3) = **20,456 ft^3**.

v) The HRT at design ave ww flow can be calculated as:
$HRT_{des\ ave}$ = reactor liquid volume*7.48/[Q*10^6/(24*60)] = 20,456*7.48/[1.5*10^6/(24*60)] = **147 min**

$HRT_{peak\ hr}$ = $HRT_{des\ ave}$/peak hour factor = 147/4 = **37 min**

vi) Calculation of the estimated effluent BOD concentration from the first stage as shown above for the single stage process gives a value of **11 mg/L**.

b) For the second stage:

i) The design value for the nitrification **SALR** is **0.63 g/m^2/d**, calculated in the same way as it was for the single stage nitrification process.

ii) The NH$_3$-N loading rate will be (1.5 MGD)(35 mg/L)(8.34 lb/MG/mg/L) = 437.9 lb/day = (437.9 lb/day)*(453.59 g/lb) = **198,604 g BOD/day**

iii) Required carrier surface area = (198,604 g/day)/(0.63 g/m²/day)
    = 316,914 m².

Required carrier volume = 316914 m²/600 m²/m³. = 528.19 m³ = (528.19 m³)(3.2808³ ft³/m³) = **18,652 ft³**.

iv) For 40% carrier fill: Required tank volume = 18652 ft³/0.40 = **46,632 ft³**.

v) The volume of liquid in the reactor can be calculated as:
tank volume − [carrier volume(1 − void %)],   Thus the volume of liquid is: 46632 − [18652(1 − 0.60)] = **42,736 ft³**.

vi) The HRT at design ave ww flow can be calculated as:
$HRT_{des\ ave}$ = reactor liquid volume*7.48/[Q*10⁶/(24*60)] = 42736*7.48/[1.5*10⁶/(24*60)] = **307 min**

$HRT_{peak\ hr}$ = $HRT_{des\ ave}$/peak hour factor = 307/4 = **77 min**

vii) Calculation of the alkalinity requirement is shown in the **Figure 30** spreadsheet screenshot below. Using the equivalent weight of $CaCO_3$ as 50, the equivalent weight of $NaHCO_3$ as 84, the alkalinity use for nitrification as 7.14 g $CaCO_3$/g NH_3-N and the target effluent alkalinity as 80 mg/L as $CaCO_3$, give the calculated alkalinity requirement as 166.3 mg/L as $CaCO_3$. The rate of alkalinity addition needed can then be calculated as: (1.5 MGD)(166.3 mg/L)*8.34 = **2080.9 lb/day as $CaCO_3$**. Multiplying this by the ratio of the equivalent weight of $NaHCO_3$ to the equivalent weight of $CaCO_3$ gives the daily $NaHCO_3$ requirement as **3495.9 lb/day $NaHCO_3$**.

MBBR Process Design Calculations - U.S. units
Two-Stage Process for BOD Removal & Nitrification

Instructions: Enter values in blue boxes. Spreadsheet calculates values in yellow boxes

I. Wastewater Parameter Inputs

1. Parameters for Both First and Second Stage

Design ww Flow Rate, Q =	1.5	MGD	Peak Hour Factor =	4	
			Prim. Effl. Alkalinity =	140	mg/L as $CaCO_3$

2. Parameters for First Stage:

Data points for SARR/SALR vs SALR

			SALR ($g/m^2/d$):	7.5	15.0
Prim. Effl. BOD, S_{o1} =	175	mg/L	SARR/SALR:	0.925	0.875

(default values above are based on the table of typical values of % BOD removal vs SALR in Worksheet 3)

Design Value of BOD Surface Area Loading Rate (SALR) =	6	$g/m^2/d$	Slope, SARR/SALR vs SALR	-0.007
See information on typical design values for SALR at the right.			Intercept, SARR/SALR vs SALR	0.975
			Est. of SARR/SALR Ratio =	0.935

(Surf. Area Removal Rate/Surf. Area Loading Rate)

3. Parameters for Second Stage:

Influent NH_4-N Conc. =	35	mg/L	Target Effl. NH_4-N Conc, NH_4-N_e =	3.3	mg/L
Min Design Temp., T =	45	°F	Click on cell H26 and on arrow to select D.O Conc.		
			D.O Conc. in Reactor =	3.0	mg/L

4. Preliminary Calculations - Design Nitrification SALR value

% NH_4-N removal =	91%		NH_4-N_e @ $SARR_{max}$ =	0.80	mg/L
Maximum SARR =	0.88	$g/m^2/d$	SARR @ NH_4-Ne, 15°C, $SARR_{15}$ =	0.88	$g/m^2/d$
			SARR @ NH_4-Ne, T °C, $SARR_T$ =	0.57	$g/m^2/d$
SARR Temp. Coeff. θ =	1.058		Design Value for nitrification $SALR$ =	0.63	$g/m^2/d$

II. Carrier Parameter and Tank Shape Inputs for both First and Second Stages

Carrier Spec. Surf. Area =	600	m^2/m^3	Click on green box and then on arrow to Select Tank Shape:	rectangular	
(value from carrier mfr/vendor)					
Liquid Depth in Tank =	8	ft	Carrier % Void Space =	60%	
Tank L:W ratio =	1.5		(from carrier mfr/vendor - only needed to calculate hydraulic detention time)		
(target L:W - only used if tank is rectangular)					

Figure 28. Screenshot of MBBR Process Design Calculations Two Stage BOD Removal and Nitrification – Part 1

III. Calculation of Carrier Volume and Required Tank Volume & Dimensions

1. First Stage Calculations - BOD Removal

(Carrier fill % is typically between 30% and 70%. Lower values are more conservative, allowing future capacity expansion or reduction of SALR by adding more carrier.

Design Carrier Fill % =	40%	(for first stage)			
BOD Daily Loading =	2189.3	lb/day			
	993022	g/day	Calculated Tank Volume =	689.6	m^3
Carrier Surf. Area needed =	165503.7	m^2	=	24352.9	ft^3
Calculated Carrier Volume =	275.839	m^3	=	182160	gal
Tank Liquid Volume =	20456.5	ft^3	Calculated Tank Width =	45.0	ft
	9740.822	24352.05	Calculated Tank Length =	67.6	ft
Estimate of BOD Surface Area			Nominal Hydraulic Retention Time at		
Removal Rate, **SARR** =	5.61	$g/m^2/d$	Design Average Flow =	147	min
Est. of BOD Removal Rate:	928475	g/day	Peak Hourly Flow =	37	min
	2046.9	lb/day	Calculated Effl BOD Conc.:	11	mg/L
			(from First Stage)		

If the calculated Effl. BOD conc. is too high, the design value of the SALR should be reduced in cell C18.

2. Second Stage Calculations - Nitrification

			Calculated Tank Volume =	1320.5	m^3
Design Carrier Fill % =	40%	(for second stage)	=	46632.1	ft^3
NH_3-N Daily Loading =	437.9	lb/day	=	348808	gal
	198604	g/day	Calculated Tank Width =	62.3	ft
Carrier Surf. Area needed =	316914.0	m^2	Calculated Tank Length =	93.5	ft
Calculated Carrier Volume =	528.190	m^3	Nominal Hydraulic Retention Time at		
Tank Liquid Volume =	42736	ft^3	Design Average Flow =	307	min
1st stage tank volume -	18652.17	46630.44	Peak Hourly Flow =	77	min
- 2nd stage tank volume =	-630.9				
			Check on BOD **SALR**:	0.2	$g/m^2/d$

To make the 2nd stage tank volume the same as the first stage, use Excel's Goal Seek process to set cell C71 equal to zero by changing the value in cell C60.

(BOD **SALR** should be < 0.5 $g/m^2/d$ in order to achieve a good nitrification rate)

Figure 29. Screenshot of MBBR Process Design Calculations Two Stage BOD Removal and Nitrification – Part 2

V. Calculation of Alkalinity Requirements

Input: Target Effluent Alkalinity = 80 mg/L

Constants needed for Calculations:

Equiv Wt. of $CaCO_3$ = 50 g/equiv.   Equiv Wt. of $NaHCO_3$ = 84 g/equiv.

Alkalinity used for Nitrification = 7.14 g $CaCO_3$/g NH_3-N

Calculations

Alkalinity to be added = 166.3 mg/L as $CaCO_3$

Daily Alkalinity Requirement = 2080.9 lb/day as $CaCO_3$

For sodium bicarbonate use to add alkalinity:

Daily $NaHCO_3$ Requirement = 3495.9 lb/day $NaHCO_3$

Figure 30. Screenshot of MBBR Process Design Calculations Two Stage BOD Removal/Nitrification – Part 3

Denitrification Background Information

In order to carry out denitrification of a wastewater flow (removal of the nitrogen from the wastewater), it is necessary to first nitrify the wastewater, that is, convert the ammonia nitrogen typically present in the influent wastewater to nitrate. Nitrification will only take place at a reasonable rate if the BOD level is quite low. Thus, an MBBR denitrification process will need a reactor for BOD removal, one for nitrification, and one for denitrification. The nitrification reactor will always follow the BOD removal reactor, because of the need for a low BOD level in the nitrification reactor. The denitrification reactor may be either before the BOD removal reactor (called pre-anoxic denitrification) or after the nitrification reactor (called post-anoxic denitrification). Flow diagrams for these two denitrification options are shown in **Figure 31** below

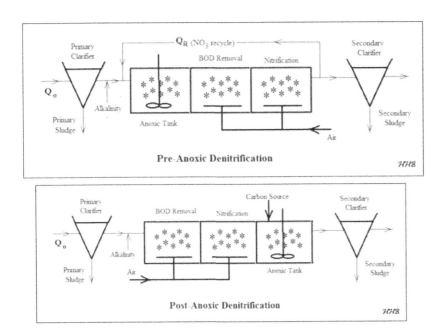

Figure 31. MBBR Denitrification Process Alternatives

A bit more information about the denitrification reactions will be useful for further discussion of these two denitrification options. The denitrification reactions, which convert nitrate ion to nitrogen gas, and hence remove it from the wastewater flow, will take place only in the absence of oxygen, that is, in an anoxic reactor. Also, the denitrification reactions require a carbon source. With these factors in mind the functioning of the pre-anoxic denitrification process and of the post-anoxic denitrification process are described in the following two paragraphs.

In a pre-anoxic denitrification process, the BOD in the primary effluent wastewater is used as the carbon source for denitrification. In this process, however the primary effluent entering the pre-anoxic reactor still has ammonia nitrogen present rather than the nitrate nitrogen needed for denitrification. A recycle flow of effluent from the nitrification reactor is used to send nitrate nitrogen to the anoxic denitrification reactor as shown in **Figure 31**.

In a post-anoxic denitrification process, the influent to the denitrification reactor comes from the nitrification reactor, so the wastewater influent ammonia nitrogen has been converted to nitrate as required for denitrification. The BOD has also been removed prior to the post anoxic denitrification reactor, however, so an external carbon source is required for the denitrification reactions. Methanol is a commonly used carbon source.

The pre-anoxic denitrification process has the advantage of not requiring an external carbon source and it reduces the BOD load to the BOD removal part of the process because BOD is used in the denitrification reactions. However, the pre-anoxic process requires an influent C/N ratio greater than 4, where C/N is taken to be BOD/TKN, and the post-anoxic process can achieve a more complete nitrogen removal.

From the Odegaard references (#5 and #6 at the end of the book) suitable criteria for determining whether to use pre- or post-anoxic denitrification are as follows:

1. Pre-anoxic denitrification is suitable if **C/N ≥ 4** and **target % Removal of N < 75%**

2. Post-anoxic denitrification should be used if **C/N < 4** or **target % Removal of N > 75%**

Post-Anoxic Denitrification Process Design Calculations

Process design of a post-anoxic denitrification MBBR system, requires sizing an MBBR tank for BOD removal, one for nitrification and one for denitrification. For all three of these reactors the key design parameter for sizing the MBBR tank is the surface area loading rate (SALR), typically with units of $g/m^2/day$, that is g/day of the parameter being removed in that reactor, coming into the MBBR tank per m^2 of carrier surface area. Using design values for wastewater flow rate and concentration of the removed parameter entering the MBBR tank, the loading rate in g/day can be calculated. Then dividing the loading rate in g/day by the SALR in $g/m^2/day$ gives the required carrier surface area in m^2. The carrier fill %, carrier specific surface area, and carrier % void space can then be used to calculate the required carrier volume, tank volume and the volume of liquid in the reactor. A typical flow diagram for a post-anoxic denitrification MBBR process is shown in **Figure 32** below.

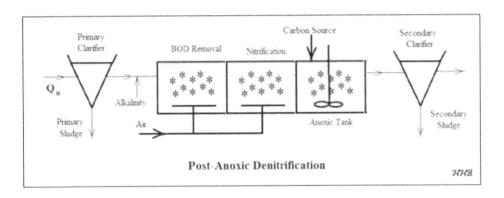

Figure 32. MBBR Flow Diagram for Post-Anoxic Denitrification

Process design calculations for the BOD removal stage and the nitrification stage will be done just as described in **Chapter 11, Two-Stage BOD Removal and Nitrification Process Design Calculations**, and illustrated with the examples in that chapter. The process design calculations for denitrification, which is the third stage in a post-anoxic denitrification process, are similar to those previously discussed and illustrated for the BOD removal stage and the nitrification stage. The graph shown in **Figure 33** (prepared using values from a similar graph in Rusten and Paulsrud's presentation in Ref #8 at the end of this book) will be used to obtain values for SARR/SALR vs SALR for the denitrification stage.

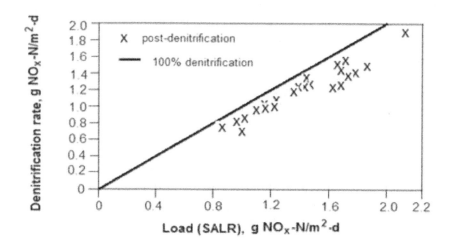

Figure 33. **SARR** vs **SALR** for Post-Anoxic Denitrification

Table 6 below shows typical SALR design values for pre-anoxic denitrification and post-anoxic denitrification in MBBR reactors. Reference #6 at the end of this book is the source for the values in **Table 5**.

Table 6. Typical Design SALR Values for Denitrification

Typical Design Values for MBBR reactors at 15°C		
Purpose	Treatment Target % Removal	Design SALR g/m^2-d
Denitrification		
Pre-DN (C/N > 4)	70 (NO_3-N)	0.90 (NO_3-N)
Post-DN (C/N > 3)	90 (NO_3-N)	2.00 (NO_3-N)

Example #16: A design wastewater flow of 1.5 MGD containing 175 mg/L BOD and 35 mg/L TKN (in the primary effluent) is to be treated in a post-anoxic denitrification MBBR process. The design SALR for the first stage is to be 6 g BOD/m^2/d. For the second stage, the SALR should be calculated for a target effluent NH$_4$-N conc. of 3.3 mg/L, min. design temperature of 45°F, and D.O. in reactor of 3.0 mg/L. For the post-anoxic stage the SALR is to be 2 g NO$_3$-N/m^2/d and the target effluent NO$_3$-N conc. is to be 5.0 mg/L.

For the third (Denitrification) stage calculate each of the following:

i) The nitrate loading
ii) The required carrier volume for a carrier with specific surface area of 600 m^2/m^3
iii) The required MBBR tank volume for a design carrier fill % of 40%
iv) The volume of liquid in the MBBR reactor for design carrier % void space of 60%.
v) The average hydraulic retention time at design average wastewater flow and at design peak hourly flow if the design peak hour factor is 4.
vi) The estimated effluent NO$_3$-N concentration from the denitrification stage.
vii) The alkalinity requirement in lb/day as CaCO$_3$ and in lb/day NaHCO$_3$, for target effluent alkalinity of 80 mg/L as CaCO$_3$.
viii) The methanol requirement in lb/day for methanol use as the carbon source.

Note that the process design calculations for the BOD removal stage and the nitrification stage of this process will be the same as those used in Example #14 in Chapter 11 for a two-stage BOD removal and nitrification process.

Solution - The solution is shown in **Figure 34**, **Figure 35**, and **Figure 36**, which are screenshots of different parts of an Excel worksheet used to carry out the calculations for this example. **Figure 34** is from the top part of the worksheet and shows the user inputs and the calculation of the estimated **SARR/SALR** ratio for the denitrification stage. **Figure 35** is from the middle of the worksheet and shows the answers for parts i) through vi), as follows:

i) The NO_3-N loading rate will be (1.5 MGD)(35 − 5 mg/L)(8.34 lb/MG/mg/L) = 396.6 lb/day = (396.6 lb/day)*(453.59 g/lb) = **179,879 g NO_3-N/day**

ii) Required carrier surface area = (179,879 g/day)/(2 g/m²/day) = 89,939 m².

Required carrier volume = 89,939 m²/600 m²/m³ = 149.90 m³

iii) For 40% carrier fill: Required tank volume = 149.90 m³/0.40 = **374.75 m³**.

iv) The volume of liquid in the reactor can be calculated as:
tank volume − [carrier volume(1 − void %)], Thus the volume of liquid is: 374.75 − [149.90(1 − 0.60)] = **314.79 m³** = 314.79*3.2808³ = **11,117 ft³**.

v) The HRT at design ave ww flow can be calculated as:
$HRT_{des\ ave}$ = reactor liquid volume*7.48/[Q*10⁶/(24*60)] = 11,117*7.48/[01.5*10⁶/(24*60)] = **80 min**

$HRT_{peak\ hr}$ = $HRT_{des\ ave}$/peak hour factor = 80/4 = **20 min**

vi) Calculation of the estimated effluent NO_3-N concentration from the second stage, as shown above for the BOD removal process, is shown in the **Figure 35** spreadsheet screenshot below and gives a value of **4.8 mg/L**,

vii) Calculation of the alkalinity requirement is shown in the **Figure 35** spreadsheet screenshot below. Using the equivalent weight of $CaCO_3$ as 50, the equivalent weight of $NaHCO_3$ as 84, the alkalinity use for nitrification as 7.14 g $CaCO_3$/g NH_3-N, the alkalinity produced by denitrification as 3.56 g $CaCO_3$/g NO_3-H, and the target effluent alkalinity as 80 mg/L as $CaCO_3$, give the calculated alkalinity requirement as 81.9 mg/L as $CaCO_3$. The rate of alkalinity addition needed can then be calculated as: (1.5 MGD)(81.9 mg/L)*8.34 = **1024.9 lb/day as $CaCO_3$**. Multiplying this by the ratio of the equivalent weight of

NaHCO$_3$ to the equivalent weight of CaCO$_3$ gives the daily NaHCO$_3$ requirement as **1721.8 lb/day NaHCO$_3$**.

viii) Calculation of the methanol requirement in lb/day is shown at the bottom of the **Figure 35** screenshot. As shown, the calculations use the constants, 4.6 lb COD/lb NO$_3$-N removed and 1.5 lb COD/lb Methanol. The required methanol dosage is then calculated as: 4.6/1.5 = 3.1 lb methanol /lb NO$_3$-N removed. The methanol requirement in lb/day is then equal to 3.1 times the previously calculated NO$_3$-N removal rate of 337.1 lb/day, or **1033.7 lb/day**.

MBBR Process Design Calculations - U.S. units
Post-Anoxic Denitrification Process

Instructions: *Enter values in blue boxes. Spreadsheet calculates values in yellow boxes*

I. Wastewater Parameter Inputs

1. Parameters for All Three Stages

Design ww Flow Rate, Q_o =	1.5	MGD	Prim. Effl. BOD, S_o =	175	mg/L
Prim. Effl. TKN Conc. =	35	mg/L	Prim. Effl. NO_3-N Conc. =	0	mg/L
Peak Hour Factor =	4		Prim. Effl. Alkalinity =	140	mg/L as $CaCO_3$

2. Parameters for First (BOD Removal) Stage:

Data points for SARR/SALR vs SALR

Design Value of BOD Surface			SALR (g/m²/d):	7.5	15.0
Area Loading Rate (SALR) =	6	g/m²/d	SARR/SALR:	0.925	0.875

See information on typical design values for SALR at right.

(default values above are from the table of typical values of % BOD removal vs SALR shown at the right from ref #2)

Est. of SARR/SALR Rato =	0.935			
(Surf. Area Removal Rate/Surf. Area Loading Rate)		Slope, SARR/SALR vs SALR:	-0.007	
		Intercept, SARR/SALR vs SALR:	0.975	

3. Parameters for Second (Nitrification) Stage:

Influent NH_4-N Conc. =	25	mg/L	Target Effl NH_4-N Conc, NH_4-N_e =	3.3	mg/L
Min Design Temp., T =	45	°F	Click on cell H26 and on arrow to select D.O Conc.		
			D.O Conc. in Reactor =	3.0	mg/L

4. Preliminary Calculations - Design Nitrification SALR value

% NH_4-N removal =	87%		NH_4-N_e @ $SARR_{max}$ =	0.80	mg/L
Maximum SARR =	0.88	g/m²/d	SARR @ NH_4-Ne, 15°C, $SARR_{15}$ =	0.88	g/m²/d
			SARR @ NH_4-Ne, T °C, $SARR_T$ =	0.57	g/m²/d
SARR Temp. Coeff, θ =	1.058		Design Value for nitrification SALR =	0.65	g/m²/d

5. Parameters for Third (Post-Anoxic) Stage:

Data points for SARR/SALR vs SALR

Target Effl. NO_3-N Conc. =	5.0	mg/L	SALR (g/m²/d):	1	2.0
Design Value of BOD Surface			SARR/SALR:	0.880	0.850
Area Loading Rate (SALR) =	2.0	g/m²/d	(default values above are from a graph of		
See information on typical design values for SALR at right.			Post-Anoxic SARR vs SALR in ref #6 below right) (The graph is shown at the right.)		
Est. of SARR/SALR Rato =	0.85		Slope, SARR/SALR vs SALR:	-0.030	
(Surf. Area Removal Rate/Surf. Area Loading Rate)			Intercept, SARR/SALR vs SALR:	0.910	

Figure 34. Screenshot – Post-Anoxic Denitrification Design Calculations – Part 1

II. Determine whether to use Pre-Anoxic or Post-Anoxic Denitrification

Carbon:Nitrogen Ratio, **C/N** = 5.0   Use this Worksheet for Post-Anoxic
Target % N removal = 86%   Denitrification calculations

III. Carrier and Tank Shape Parameter Inputs for all Three Stages

Carrier Spec. Surf. Area =	600	m^2/m^3
(from carrier mfr/vendor)		
Liquid Depth in Tank =	8	ft
Tank L:W ratio =	1.5	
(target L:W - only used if tank is rectangular)		

Click on green box and then on arrow to Select Tank Shape: **rectangular**

Carrier % Void Space = 60%
(from carrier mfr/vendor - only needed to calculate hydraulic detention time)

IV. Calculation of Carrier Volume and Required Tank Volume & Dimensions

1. First Stage (BOD Removal) Calculations

(Carrier fill % is typically between 30% and 70%. Lower values are more conservative, allowing future capacity expansion or reduction of SALR by adding more carrier.

Design Carrier Fill % =	60%	(for first stage)	Required Tank Volume =	459.7	m^3
BOD Daily Loading =	2189.3	lb/day		16235.3	ft^3
BOD Daily Loading =	993022	g/day		121440	gal
Carrier Surf. Area needed =	165504	m^2			
Calculated Carrier Volume =	275.84	m^3	Calculated Tank Width =	36.8	ft
Tank Liquid Volume =	12339.0	ft^3	Calculated Tank Length =	55.2	ft

Nominal Hydraulic Retention Time at

Estimate of BOD Surface Area			Design Average Flow =	8.9	min
Removal Rate, **SARR** =	5.61	$g/m^2/d$	Peak Hourly Flow =	2.2	min
Est. of BOD Removal Rate:	928475	g/day	Calculated Effl BOD Conc.:	11	mg/L
	2046.9	lb/day	If the calculated Effl. BOD conc. is too high, the		

design value of SALR (in cell C17) should be reduced.

Figure 35. Screenshot – Post-Anoxic-Denitrification Design Calculations – Part 2

VII. Calculation of Alkalinity Requirements

Input: Target Effluent Alkalinity = 80 mg/L as $CaCO_3$

Constants needed for Calculations:

Equiv Wt. of $CaCO_3$ = 50 g/equiv.   Equiv Wt. of $NaHCO_3$ = 84 g/equiv.

Alkalinity used for Nitrification = 7.14 g $CaCO_3$/g NH_3-N removed
Alkalinity produced by Denitrification = 3.57 g $CaCO_3$/g NO_3-N removed

Calculations

Alkalinity to be added = 81.9 mg/L as $CaCO_3$
Daily Alkalinity Requirement = 1024.9 lb/day as $CaCO_3$

For sodium bicarbonate use to add alkalinity:

Daily $NaHCO_3$ Requirement = 1721.8 lb/day $NaHCO_3$

VII. Calculation of Carbon Source Requirements

Inputs:
Carbon Source to be used: Methanol
COD Requirement for Denitrification = 4.6 lb COD/lb NO_3-N removed
COD Content of Carbon Source = 1.5 lb COD/lb Carbon Source

Calculations

Carbon Source Dosage = 3.1 lb Carbon Source/lb NO_3-N removed
Daily Carbon Source Requirement = 1033.7 lb/day

Figure 36. Screenshot – Post-Anoxic Denitrification Design Calculations – Part 3

Pre-Anoxic Denitrification Process Design Calculations

The process design calculations for pre-anoxic denitrification, are similar to those just discussed for a post-anoxic denitrification process. The graph shown in **Figure 37** (prepared using values from a similar graph in Rusten and Paulsrud's presentation in Ref #8 at the end of this book) will be used to obtain values for SARR/SALR vs SALR for the pre-anoxic denitrification stage

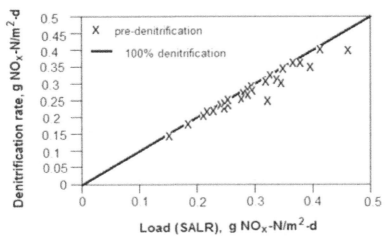

Figure 37. SARR vs SALR for Pre-Anoxic Denitrification

A typical flow diagram for a pre-anoxic denitrification MBBR process is shown in **Figure 38** below. As discussed previously for a post-anoxic denitrification MBBR system, process design of a pre-anoxic denitrification MBBR system also requires sizing an MBBR tank for BOD removal, one for nitrification and one for denitrification. Process design for the BOD removal and nitrification stages is essentially the same as just discussed for the post-anoxic denitrification process. The primary difference from the post-anoxic denitrification process design calculations is for the denitrification stage, which will be discussed and illustrated with example calculations below.

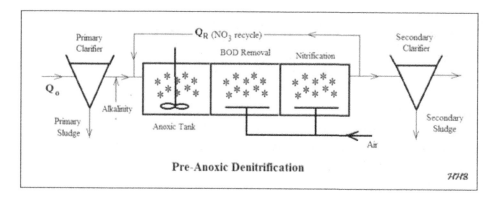

Figure 38. MBBR Flow Diagram for Pre-Anoxic Denitrification

Example #17: Carry out the process design as described below for the denitrification stage of a pre-anoxic denitrification process with the wastewater flow and concentrations given in the previous Examples. [1.5 MGD containing 175 mg/L BOD and 35 mg/L TKN (in the primary effluent)]. Consider that the primary effluent alkalinity is 140 mg/L as $CaCO_3$ and the design SALR for the denitrification stage is to be 0.9 g NO_3-N/m²/d.

For the first stage (Denitrification) calculate each of the following:

i) The nitrate loading
ii) The required carrier volume for a carrier with specific surface area of 600 m²/m³
iii) The required MBBR tank volume for a design carrier fill % of 40%
iv) The volume of liquid in the MBBR reactor for design carrier % void space of 60%.
v) The average hydraulic retention time at design average wastewater flow and at design peak hourly flow if the design peak hour factor is 4.
vi) The required NO_3-N recycle rate in order to achieve a target NO_3-N concentration of 9 mg/L.
vii) The alkalinity requirement in lb/day as $CaCO_3$ and in lb/day $NaHCO_3$, for target effluent alkalinity of 80 mg/L as $CaCO_3$.

Solution - The solution is shown in **Figure 39, Figure 40**, and **Figure 41**, which are screenshots of different parts of an Excel worksheet used to carry out the calculations for this example. **Figure 39** is from the top part of the worksheet and shows the user inputs and the calculation of the estimated **SARR/SALR** ratio for the denitrification stage. Note also that a user input value is needed for the estimated NO_3-N recycle ratio. This initial estimated value is used in an iterative calculation to determine the required NO_3-N recycle ratio in order to achieve the target effluent NO_3-N concentration.

Figure 40 is from the middle of the worksheet and shows the answers for parts i) through vi). **Figure 41** is from the bottom of the worksheet and shows the Alkalinity calculations which answer part vii). The calculations and results are as follows:

i) Most of the nitrate loading to the pre-anoxic denitrification tank is typically in the NO_3-N recycle flow rather than in the primary effluent flow entering the tank. The NO_3-N loading rate will be (1.5 MGD)(Prim Effl NO_3-N)(8.34 lb/MG/mg/L) + (1.5 MGD)(Recycle Ratio)(Target Effl NO_3-N)(8.34) = 297.0 lb/day = (297.0 lb/day)*(453.59 g/lb) = **134,716 g NO_3-N/day**

ii) Required carrier surface area = (134,716 g/day)/(0.9 g/m²/day) = 149,685 m².

Required carrier volume = 145,744 m²/600 m²/m³. = 249.47 m³

iii) For 40% carrier fill: Required tank volume = 249.47 m³/0.40 = **623.7 m³**.

iv) The volume of liquid in the reactor can be calculated as:
tank volume – [carrier volume(1 – void %)], Thus the volume of liquid is: 623.7 – [249.47(1 – 0.60)] = **523.9 m³** = 523.9(3.2808³) = **18,501 ft³**.

v) The HRT at design ave ww flow can be calculated as:
$HRT_{des\ ave}$ = reactor liquid volume*7.48/[Q*10⁶/(24*60)] = 18501*7.48/[1.5*10⁶/(24*60)] = **133 min**

$$HRT_{peak\ hr} = HRT_{des\ ave}/\text{peak hour factor} = 133/4 = \mathbf{\underline{33\ min}}$$

vi) The required NO_3-N recycle ratio is calculated with the iterative process described in blue at the bottom of **Figure 40**. For this iterative process, the NO_3-N removal rate is calculated two different ways, one using the estimated SARR and the carrier surface area and the other using the wastewater flow rate times the influent TKN concentration minus the sum of the effluent nitrate and ammonia nitrogen. Excel's Goal Seek process is then used to set the difference between the two different calculations equal to zero by changing the estimated value of the NO_3-N recycle ratio. This process results in the required NO_3-N recycle ratio calculated to be **2.64**.

vii) Calculation of the alkalinity requirement is shown in the **Figure 41** spreadsheet screenshot. Using the equivalent weight of $CaCO_3$ as 50, the equivalent weight of $NaHCO_3$ as 84, the alkalinity use for nitrification as 7.14 g $CaCO_3$/g NH_3-N, the alkalinity produced by denitrification as 3.57 g $CaCO_3$/g NO_3-H, and the target effluent alkalinity as 80 mg/L as $CaCO_3$, give the calculated alkalinity requirement as 97.1 mg/L as $CaCO_3$. The rate of alkalinity addition needed can then be calculated as: (1.5 MGD)(91.1 mg/L)*8.34 = **1214.5 lb/day as $CaCO_3$**. Multiplying this by the ratio of the equivalent weight of $NaHCO_3$ to the equivalent weight of $CaCO_3$ gives the daily $NaHCO_3$ requirement as **2040.3 lb/day $NaHCO_3$**.

MBBR Process Design Calculations - U.S. units
Pre-Anoxic Denitrification Process

Instructions: Enter values in blue boxes. Spreadsheet calculates values in yellow boxes

I. Wastewater Parameter Inputs

1. Parameters for All Three Stages

Design ww Flow Rate, Q_o =	1.5	MGD	Prim. Effl. BOD, S_o =	175	mg/L
Prim. Effl. TKN Conc. =	35	mg/L	Prim. Effl. NO_3-N Conc. =	0	mg/L
Peak Hour Factor =	4		Prim. Effl. Alkalinity =	140	mg/L as $CaCO_3$

2. Parameters for First (Pre-Anoxic) Stage:

Data points for SARR/SALR vs SALR

Target Effl. NO_3-N Conc. =	9	mg/L	SALR (g/m²/d):	0.2	0.5
Est. of NO_3-N Recycle Ratio	2.64		SARR/SALR:	0.95	0.94

(Q_R/Q_o) - An estimate is needed here to start (default values above are from a graph of Pre-Anoxic
the iterative calculation in Sec IV below   SARR vs SALR in ref #6 below right)

Design Value of NO_3-N Surface   (The graph is shown at the right.)

Area Loading Rate (SALR) =	0.9	g/m²/d	Slope, SARR/SALR vs SALR:	-0.033
See information on typical design			Intercept, SARR/SALR vs SALR:	0.957
values for SALR at right.			Est. of SARR/SALR Ratio =	0.927

(Surf. Area Removal Rate/Surf. Area Loading Rate)

3. Parameters for Second (BOD Removal) Stage:

Data points for SARR/SALR vs SALR

Design Value of BOD Surface			SALR (g/m²/d):	7.5	15.0
Area Loading Rate (SALR) =	6	g/m²/d	SARR/SALR:	0.925	0.875

See information on typical design   (default values above are from the table of typical values
values for SALR at right.   of % BOD removal vs SALR shown at the right from ref #2)

Est. of SARR/SALR Ratio =	0.935		Slope, SARR/SALR vs SALR:	-0.007
(Surf. Area Removal Rate/Surf. Area Loading Rate)			Intercept, SARR/SALR vs SALR:	0.975

4. Parameters for Third (Nitrification) Stage:

Min Design Temp., T =	45	°F	Target Effl NH_4-N Conc, NH_4-N_e =	4	mg/L
			Click on cell H26 and on arrow to select D.O Conc.		
			D.O Conc. in Reactor =	3.0	mg/L

5. Preliminary Calculations - Design SALR value for Nitrification

% NH_4-N removal =	89%		NH_4-N_e @ $SARR_{max}$ =	0.80	mg/L
Maximum SARR =	0.88	g/m²/d	SARR @ NH_4-Ne, 15°C, $SARR_{15}$ =	0.88	g/m²/d
			SARR @ NH_4-Ne, T °C, $SARR_T$ =	0.57	g/m²/d
SARR Temp. Coeff, θ =	1.058		Design Value for nitrification SALR =	0.64	g/m²/d

Figure 39. Screenshot – Pre-Anoxic Denitrification Design Calculations – Part 1

II. Determine whether to use Pre-Anoxic or Post-Anoxic Denitrification

Carbon:Nitrogen Ratio, C/N = 5.0
Target % N removal = 74%

Use this Worksheet for Pre-Anoxic Denitrification calculations

III. Carrier and Tank Shape Parameter Inputs for all Three Stages

Carrier Spec. Surf. Area = 600 m^2/m^3
(from carrier mfr/vendor)

Click on green box and then on arrow to Select Tank Shape: **rectangular**

Liquid Depth in Tank = 8 ft
Tank L:W ratio = 1.5
(target L:W - only used if tank is rectangular)

Carrier % Void Space = 60%
(from carrier mfr/vendor - only needed to calculate hydraulic detention time)

IV. Calculation of Carrier Volume and Required Tank Volume & Dimensions

1. First Stage (Pre-Anoxic Tank) Calculations

(Carrier fill % is typically between 30% and 70%. Lower values are more conservative, allowing future capacity expansion or reduction of SALR by adding more carrier.

Design Carrier Fill % =	40%	(for first stage)	Calculated Tank Volume =	623.7	m^3
NO_3-N Daily Loading =	297.0	lb/day		22025.3	ft^3
NO_3-N Daily Loading =	134716.2	g/day		164749	gal
Carrier Surf. Area needed =	149684.7	m^2	Calculated Tank Width =	42.8	ft
Calculated Carrier Volume =	249.475	m^3	Calculated Tank Length =	64.3	ft
Tank Liquid Volume =	18501.2	ft^3	Nominal Hydraulic Retention Time at		
Estimate of NO_3-N Surface Area			Design Average Flow =	133	min
Removal Rate, **SARR** =	0.83	$g/m^2/d$	Peak Hourly Flow =	33	min
Req'd NO_3-N Removal Rate:	275.22	lb/day	Est. of NO_3-N Removal Rate:	275.22	g/day
(for Effl. NO_3-N = Target Value)			Est. Rem Rate - Req'd Rate:	0.0000	g/day

NOTE: Use Excel's "Goal Seek" to find Q_R/Q_o as follows: Place the cursor on cell H73 and click on "Goal Seek" (in the "tools" menu of older versions and under "Data - What if Analysis" in newer versions of Excel). Enter values to "Set cell:" H73, "To value:" 0, "By changing cell:" c17, and click on "OK". The calculated value of Q_R/Q_o will appear in cell C17 and cell H73 should equal zero if the process worked properly. Note that an initial estimate of Q_R/Q_o is needed in cell C17 to start

Figure 40. Screenshot – Pre-Anoxic Denitrification Design Calculations – Part 2

VII. Calculation of Alkalinity Requirements

Input: Target Effluent Alkalinity = [80]   mg/L

Constants needed for Calculations:

Equiv Wt. of $CaCO_3$ = [50] g/equiv.   Equiv Wt. of $NaHCO_3$ = [84] g/equiv.

Alkalinity used for Nitrification = [7.14] g $CaCO_3$/g NH_3-N removed
Alkalinity produced by Denitrification = [3.57] g $CaCO_3$/g NO_3-N removed

Calculations

Alkalinity to be added = [97.1] mg/L as $CaCO_3$
Daily Alkalinity Requirement = [1214.5] lb/day as $CaCO_3$

For sodium bicarbonate use to add alkalinity:

Daily $NaHCO_3$ Requirement = [2040.3] lb/day $NaHCO_3$

Figure 41. Screenshot – Pre-Anoxic Denitrification Design Calculations – Part 3

Description of the MBR (Membrane Bioreactor) Process

Initial Development of the MBR Process

Membrane filtration has been used for quite some time in a variety of ways in water and wastewater treatment. It was not until the 1970's, however that Research at Rensselaer Polytechnic Institute proposed coupling the activated sludge process with membrane filtration. Dorr-Oliver introduced the first membrane bioreactor (MBR) wastewater treatment processes with a flat sheet ultrafiltration plate and frame membrane. This process was put into use in Japan in the 1970s and 1980s, but did not come into very widespread use around the world. In 1989 Yamamoto et al (Ref #11 at the end of this book) developed and introduced submerged membranes with a membrane module placed directly in the aeration tank rather than as a separate process following the aeration tank. This innovation accelerated interest in, development of, and spreading use of MBR wastewater treatment processes. Currently almost all MBR processes being installed have a submerged membrane module, as shown in **Figure 41** below.

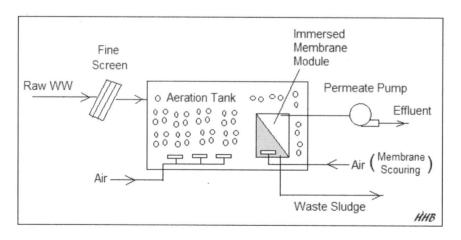

Figure 41. MBR Wastewater Treatment Typical Process Flow Diagram

Comparison of MBR Process with Conventional Activated Sludge

The first difference to note is that the use of membrane filtration allows an MBR process to produce a significantly higher quality effluent than that obtainable from a conventional activated sludge process. Secondly, the mixed liquor suspended solids concentration (MLSS) and the solids residence time (SRT) are limited in a conventional activated sludge process by the need to produce a sludge with good settling characteristics in the secondary clarifier. This is not a requirement for MBR processes, because the final effluent is produced by filtration rather than by sedimentation. Thus, the MLSS and SRT can both be larger for MBR processes than the typical values used for conventional activated sludge. This results in a smaller aeration tank volume needed for an MBR process than that needed for conventional activated sludge treating the same flow.

On the other hand, the need to keep the membranes from getting fouled increases operating costs over typical values for conventional activated sludge. The method that has evolved for cleaning the membranes and keeping them from getting fouled is a fairly high rate of aeration below the membrane module, so that the air bubbles keep the membrane clean. This results in a higher aeration cost than that typically required for conventional activated sludge.

Pretreatment Prior to the MBR Aeration Tank

Primary clarification is not typically used before the MBR aeration tank, however, coarse screening and grit removal should be used if warranted by the wastewater characteristics, and fine screening (2 – 3 mm pore size) should be used as the final pretreatment step before the aeration tank.

MBR Process Alternatives

The MBR process alternatives are similar to those for a conventional activated sludge process. An MBR process may be used for the following:

1. BOD removal and nitrification

2. BOD removal and nitrification together with Pre-anoxic denitrification

Post-anoxic denitrification does not work too well with an MBR process, because the membrane module must be at the effluent end of the process but cannot be submerged in the post-anoxic tank, because scouring aeration is required to keep the membrane from fouling.

Overview of MBR Process Design Calculations

The membrane surface area requirement, volume requirement, and required scouring air flow rate for the membrane module can be calculated using membrane module properties typically available from membrane module manufacturers or vendors together with wastewater flow information.

Sizing of the aeration tank will be done in a manner similar to that used for a completely mixed activated sludge process.

Membrane Module Process Design Calculations

The membrane module properties typically used for process design and their units are: 1) Average Membrane Flux, **J**, in L/hr/m^2, 2) Module Packing Density, ϕ, in m^2/m^3, and 3) Specific Aeration Demand, **SAD$_M$** in m^3 air/hr/m^2 membrane. Using these membrane and membrane module properties, the required membrane area, membrane module volume, and scouring air flow rate can be calculated as follows:

Membrane area, **A$_M$ = Q$_o$/J**, where **Q$_o$** is the wastewater flow rate in L/hr. With **Q$_o$** in m^3/day, the equation with conversion factors becomes: **A$_m$ = (Q$_o$/24)*1000/J**. With **Q$_o$** in MGD, the equation with conversion factors becomes: **A$_m$ = (Q$_o$/24)*1000000/[(J/3.7854)/(3.2808^2)]**. Note that in this case, the flow rate has been converted to gal/day and **J** has been converted to gal/hr/ft^2.

The membrane module volume is then calculated as: **V$_M$ = A$_M$/ϕ**. For S.I. units no conversion factors are needed. For U.S. units, ϕ must be converted from m^2/m^3 to ft^2/ft^3.

The scouring air flow required to keep the membrane from getting fouled is calculated as: Required Scouring Air Flow = **SAD$_M$*A$_M$/60**. This gives the required scouring air flow in m^3/min. To calculate the required scouring air flow in cfm, SADM should be converted to ft^3 air/hr/ft^2 membrane.

Example #18: a) A design wastewater flow 7571 m^3/d is to be treated with an MBR wastewater treatment system. The design membrane module properties are: average membrane flux, **J** = 12 L/hr/m^2 ; module packing density, ϕ = 120 m^2/m^3; specific aeration demand, **SAD$_M$** = 0.3 m^3 air/hr/m^2 membrane. Calculate the required membrane area, membrane module volume, and scouring air flowrate.

b) Repeat for a design wastewater flow of 1.5 MGD with the same membrane module properties.

Solution: a) **Figure 42** below is a screenshot of an Excel spreadsheet set up to make the membrane module calculations described above for part a), with S.I. units. The given membrane module properties were entered in the blue cells. A_m, V_m and scouring air flow are calculated by the worksheet in the yellow cells, using the equations given above. As shown in **Figure 42**, the results are: $A_m = 26,288$ m^2, $V_m = 219$ m^3, and scouring air flow required = 131 m^3/min.

b) **Figure 43** is a similar screenshot with the calculations in U.S. units, as described above. The results in U.S. units are: $A_m = 212,217$ ft^2, $V_m = 5802$ ft^3, and scouring air flow required = 3481 cfm.

```
MBR Process Design Calculations - S.I. units
    Membrane Module Sizing Calculations

Instructions: Enter values in blue boxes. Spreadsheet calculates values in yellow boxes

1. User Inputs, Membrane/Membrane Module Parameters:
        (values typically available from membrane manufacturer or vendor)

Ave. Membrane Flux, J =    [ 12 ]   L/hr/m²      Spec. Aer. Demand, SAD_M =  [ 0.3 ]
Module packing                                          (m³ air/hr/m² membrane)
    density, φ =           [ 120 ]  m²/m³

2. Process Design Calculations  (done by worksheet)

Membrane Area, A_M =       [ 26,288 ]  m²      Membrane Module Vol., V_M =  [ 219 ]  m³

Scouring Air Flow Required:  [ 131 ]  m³/min  (ACMM)  =  [ 135 ]  SCMM

(This is the scouring air flow rate needed for the membrane module,
    typically provided by a coarse bubble diffuser system.)
```

Figure 42. Excel Worksheet with Membrane Module Calculations
S.I. units

MBR Process Design Calculations - U.S. units
Membrane Module Sizing Calculations

Instructions: *Enter values in blue boxes. Spreadsheet calculates values in yellow boxes*

1. User Inputs, Membrane/Membrane Module Parameters:
(values typically available from membrane manufacturer or vendor)

Ave. Membrane Flux, J =	12	$L/hr/m^2$ =	0.295	$gal/hr/ft^2$	
Module packing density, ϕ =	120	m^2/m^3 =	36.6	ft^2/ft^3	
Spec. Aer. Demand. SAD_M =	0.3	m^3 air/hr/m^2 membrane =	0.984		
			(ft^3 air/hr/ft^2 membrane)		

2. Process Design Calculations (done by worksheet)

Membrane Area, A_M = 212,217 ft^2 Membrane Module Vol., V_M = 5,802 ft^3

Scouring Air Flow Required: 3481 ft^3/min (ACFM) = 3576 SCFM

(This is the scouring air flow rate needed for the membrane module, typically provided by a coarse bubble diffuser system.)

Figure 43. Excel Worksheet with Membrane Module Calculations U.S. units

Process Design Calculations for BOD Removal and Nitrification

The aeration tank will be essentially a completely mixed tank, so the process design calculations for the aeration tank can be done using a completely mixed activated sludge (CMAS) process design procedure. The calculations described in this section and illustrated with example calculations follow the CMAS process design procedure presented in Metcalf & Eddy's 4th edition of *Wastewater Engineering, Treatment and Reuse*, which is the second reference in the list at the end of this book.

Required User Inputs

Quite a few user inputs are needed for the Metcalf & Eddy CMAS process design procedure. **Figure 44** below, is an Excel spreadsheet screenshot showing the wastewater parameters/characteristics inputs needed.

MBR Process Design Calculations - U.S. units
User Inputs and Constants

Instructions: *Enter values in blue boxes. Spreadsheet calculates values in yellow boxes*

User Inputs - Wastewater Parameters/Characteristics

Parameter	Value	Units	Parameter	Value	Units
Design ww Flow Rate, Q_o =	1.5	MGD	Influent TSS, TSS_o =	175	mg/L
Influent BOD, BOD_o =	175	mg/L	Influent VSS, VSS_o =	128	mg/L
sBOD, $sBOD_o$ =	115	mg/L			
			Influent TKN, TKN_o =	35	mg/L
Influent COD, COD_o =	350	mg/L	TKN peak/ave factor, FS =	1.5	
sCOD, $sCOD_o$ =	180	mg/L	Influent NH_4-N, NH_4-N_o =	25	mg/L
rbCOD, $rbCOD_o$ =	90	mg/L	Influent Alkalinity, Alk_o =	140	mg/L as $CaCO_3$
ratio, bCOD/BOD =	1.6		Aeration WW Temp., T_{ww} =	54	°F

Figure 44. Wastewater Parameter/Characteristics Inputs – U.S. units

Figure 45 below is another screenshot showing the biological kinetic coefficients for BOD removal and for nitrification that are needed, along with a couple of constants. The values shown in **Figure 44** and **Figure 45** will be used in the example calculations for MBR reactors.

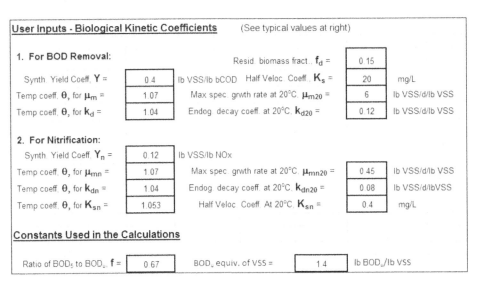

Figure 45. Inputs for Biological Kinetic Coefficients and Constants
U.S. units

Figure 46, on the next page, is a table showing ranges and typical values for the BOD removal biological kinetic coefficients. **Figure 47** is a similar table for nitrification biological kinetic coefficients.

Activated Sludge Kinetic Coefficients for heterotrophic bacteria at 20°C (for BOD removal)			
Coefficient	Unit	Range	Typical Value
μ_m	kg VSS/day/kg VSS	3.0 - 13.2	6.0
K_s	mg/L bCOD	5.0 - 40.0	20.0
Y	kg VSS/day/kg bCOD	0.30 - 0.50	0.40
k_d	kg VSS/day/kg VSS	0.06 - 0.20	0.12
f_d	dimensionless	0.08 - 0.20	0.15
θ values for temperature corrections			
for μ_m	dimensionless	1.03 - 1.08	1.07
for k_d	dimensionless	1.03 - 1.08	1.04
for K_s	dimensionless	1.00	1.00

Adapted from: Metcalf & Eddy, Inc, (Revised by Tchobanoglous, G, Burton, F.L., Stensel, H.D., *Wastewater Engineering, Treatment and Reuse*, 4th Ed., New York, NY, 2003.

Figure 46. Ranges and Typical Values for Biological Kinetic Coefficients BOD Removal

Activated Sludge Nitrification Kinetic Coefficients at 20°C			
Coefficient	Unit	Range	Typical Value
μ_{mn}	kg VSS/day/kg VSS	0.2 - 0.90	0.75
K_n	mg/L NH$_4$-N	0.50 - 1.0	0.74
Y_n	kg VSS/kg NH$_4$-N	0.10 - 0.15	0.12
k_{dn}	kg VSS/day/kg VSS	0.05 - 0.15	0.08
K_o	mg/L	0.40 - 0.60	0.50
θ values for temperature corrections			
for μ_{mn}	dimensionless	1.06 - 1.123	1.07
for k_{dn}	dimensionless	1.03 - 1.08	1.04
for K_n	dimensionless	1.03 - 1.123	1.053

Adapted from: Metcalf & Eddy, Inc. (Revised by Tchobanoglous, G, Burton, F.L., Stensel, H.D., *Wastewater Engineering, Treatment and Reuse*, 4th Ed., New York, NY, 2003.

Figure 47. Ranges and Typical Values for Biological Kinetic Coefficients Nitrification

Some additional user inputs are needed for the BOD removal/nitrification process design calculations, as shown in the **Figure 48** Excel spreadsheet screenshot.

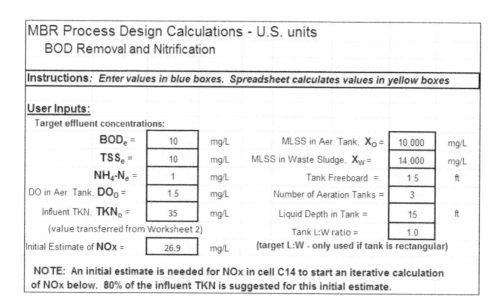

Figure 48. User Inputs for MBR Process Design Calculations BOD Removal and Nitrification

Figure 49 shows the results of the first two steps of the process design calculations made by the spreadsheet based on the user inputs shown previously and those shown in **Figure 48**. Those first two steps are calculation of the design SRT and calculation of the biomass production rate. The equations used for those calculations are as shown below. These calculations are the same for either U.S. units or S.I. units.

For calculation of Design SRT:

μ_{nm} at T_{ww} = $\mu_{nm,20}\, \theta^{T_{ww}-20}$ K_n at T_{ww} = $K_{n,20}\, \theta^{T_{ww}-20}$

k_{dn} at T_{ww} = $k_{dn,20}\, \theta^{T_{ww}-20}$ (T_{ww} must be in °C for these calculations)

μ_n = $[\, \mu_{nm} N/(K_n + N)\,]\,[\, DO/(K_o + DO)\,]$ - k_{dn} Theoretical
SRT = $1/\mu_n$

Design SRT = (FS)(Theoretical SRT)   FS = $TKN_{peak}/TKN_{average}$

For calculation of Biomass Production Rate:

μ_m at T_{ww} = $\mu_{m,20}\,\theta^{T_{ww}-20}$ k_d at T_{ww} = $k_{d,20}\,\theta^{T_{ww}-20}$

$S = K_s[1 + (k_d)SRT]/[SRT(\mu_m - k_d) - 1]$

$P_{X,bio} = QY(S_o - S)(8.34)/(1 + k_d SRT) + f_d k_d QY(S_o - S)(8.34)SRT/(1 + k_d SRT) + QY_n(NO_x)(8.34)/(1 + k_{dn} SRT)$

Note that the equation for $P_{X,bio}$, as shown above, is for use with U.S. units (Q in MGD). For S.I. units with Q in m³/d, the 8.34 conversion factor [(lb/MG}/(mg/L)] should be replaced with (1/1000) to convert g/day to kg/day.

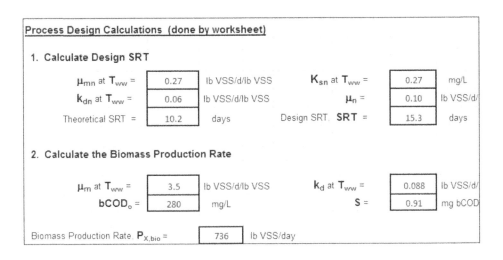

Figure 49. Calculation of Design SRT and Biomass Production Rate

Figure 50 shows the calculation of the amount of nitrogen oxidized to nitrate (NO_x), using the following equation:

$$NO_x = TKN - N_e - 0.12 P_{X,bio}/Q$$

The iterative procedure described in blue sets the difference between the estimated and calculated values of NOx to zero by changing the initial estimated value in cell C14 (shown in **Figure 8**). These calculations are the same for either U.S. or S.I. units.

> **3. Determine the Amount of Nitrogen Oxidized to Nitrate**
>
> Calculated amount of nitrogen oxidized to nitrate, **NOx** = 26.9 mg/L
>
> Difference between estimated and calculated values for **NOx** = 0.000 mg/L
> **Goal Seek Result:**
> Amount of nitrogen oxidized to nitrate, **NOx** = 26.9 mg/L
>
> NOTE: This is an iterative solution. You must use Excel's "Goal Seek" to find the NOx value as follows: Place the cursor on cell G36 and click on "goal seek" (in the "tools" menu of older versions and under "Data What If Analysis" in newer versions of Excel). Enter values to "Set cell:" G36, "To value:" 0, "By changing cell:" C14, and click on "OK". The calculated value of NOx will appear in cell E38 and cell G36 should be zero if the process worked properly. Note that cell C14 needs an initial estimate for NOx in order for the iterative solution to work properly.

Figure 50. Calculation of Amount of Nitrogen Oxidized to Nitrate

Figure 51 shows the calculation of the production rate and mass of VSS and TSS in the aeration tank, using the following equations:

bpCOD/pCOD = (BODo - sBODo)/(CODo - sCODo)

nbVSS = [1 - (bpCOD/pCOD)VSSo]

$P_{X,VSS}$ = $P_{X,bio}$ + Q(nbVSS*8.34)

$P_{X,TSS}$ = $P_{X,bio}$/(VSSo/TSSo) + Q(nbVSS*8.34) + Q(TSS$_o$ - VSS$_o$)8.34

Note that the equations for $P_{X,VSS}$ and $P_{X,TSS}$, as shown above, are for use with U.S. units (Q in MGD). For S.I. units with Q in m^3/d, the 8.34 conversion factor [(lb/MG}/(mg/L)] should be replaced with (1/1000) to convert g/day to kg/day.

Mass of MLVSS = ($P_{X,VSS}$) SRT   Mass of MLSS = ($P_{X,TSS}$) SRT

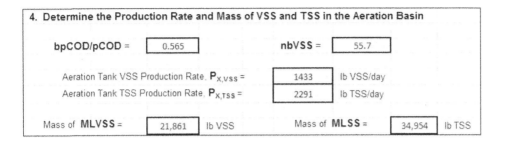

Figure 51. Calculation of Production Rate and Mass of VSS and TSS in Aeration Tank

Figure 52 shows the calculation of the aeration tank volume and dimensions, the detention time and the MLVSS concentration, using the following equations:

V = [Mass of MLSS/(MLSS*8.34)]*(1,000,000/7.48)   (for V in ft³ and Mass of MLSS in lb)

V = Mass of MLSS*1000/MLSS   (for V in m³ and Mass of MLSS in kg)

detention time: τ = **V/Q**   The actual equation used to calculate the detention time in U.S. units with **V** in ft³ and **Q** in MGD is: τ = **[(V*7.48/1,000,000)*(N*24)]/Q**   where **V** is the actual aeration tank volume for each tank and **N** is the number of aeration tanks.

For S.I. units with V in m³ and Q in m³/d, the equation is:
τ = **[V*(N*24)]/Q**

MLVSS = MLSS*(Mass of MLVSS/Mass of MLSS)

Also, the equation for the volume of a rectangular tank or the volume of a cylindrical tank are used (along with the user specified depth of water in the tank (and the L:W ratio if it is rectangular) to calculate the length and width or diameter of the aeration tank

5. Calculate Aeration Tank Volume and dimensions, Detention Time, and MLVSS (User Input needed in Blue Cells)							
Req. Aeration Vol., V_{aer} =	56,030	ft^3	Click on green box and then on arrow to Select Tank Shape:		rectangular	
Aer. + Membr. Vol., V_{tot} =	61,833	ft^3			
Req. Vol. per tank, V_{tank} =	20,611	ft^3	Actual Tank Width:		41.0	ft
Calculated Tank Width =	37.1	ft	Actual Tank Length:		41.0	ft
Calculated Tank Length =	37.1	ft	Actual Tank Aeration Volume =		23281	ft^3
Tank Wall Height =	16.5	ft	Membrane Module Vol. per Tank =		1934	ft^3
Aeration Det'n time, τ =	8.36	hr	MLVSS =		6254	mg/L

Figure 52. Calc. of Aeration Tank Volume and Dimensions, Detention Time, & MLVSS

Figure 53 shows the calculation of the F/M ratio, the Volumetric BOD loading, and the sludge wasting rate, using the following equations:

F/M = QS_o/(V*MLVSS)

This equation can be used as it is for S.I. units with Q in m^3/day and V is the total volume of all aeration tanks in m^3. For U.S. units with Q in MGD and V in ft3, some conversion factors are needed, giving the following equation:

F/M = QS_o/[(V*7.48/1,000,000)*MLVSS]

For S.I. units with Q in m^3/d and V in m^3, the equation for volumetric BOD loading is:

Vol BOD loading = QS_o(1/1000)/V NOTE: V is the total vol. of all aeration tanks.

For U.S. units with Q in MGD and V = total aeration vol. in ft^3, the volumetric BOD loading in lb BOD/d/1000 ft^3 can be calculated with the following equation:

Vol BOD loading = QS_o*8.34/(V/1000)/V

For S.I. units with total aeration tank volume, **V**, in m³ the sludge wasting rate, **Qw** can be calculated in m³/d using the following equation with **TSS_W** being the TSS concentration in the wasted sludge:

$$Q_w = [(V*MLSS/SRT*TSS\_W)]$$

For U.S. units with total aeration tank volume, **V**, in ft³ the sludge wasting rate, **Qw**, can be calculated in gal/d using the following equation:

$$Q_w = [(V*7.48*MLSS/SRT*TSS\_W)]$$

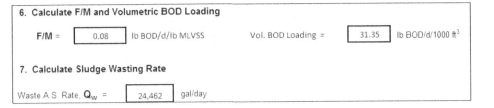

Figure 53. Calculation of F/M, Volumetric BOD loading, and Sludge Wasting Rate

Oxygen/Air Requirement and Blower Calculations for BOD removal and nitrification can be made as described in **Chapter 7**, for the activated sludge process. **Figure 54** shows the S.I. version of user inputs needed to make the oxygen/air/blower calculations using the "Rule of Thumb" guidelines presented in **Chapter 7**. **Figure 55** shows the U.S. version.

8. Oxygen/Air Requirement and Blower Calculations (for the Aeration Tank)

i) INPUTS (Values of "Rule of Thumb" Constants for the Calculations - See info at right)

O_2 needed per kg BOD =	1.50	kg O_2/kg BOD	Depth of Diffusers =	4.4	m
O_2 needed per kg NH_4-N =	4.57	kg O_2/kg NH_4-N	Standard Temperature =	20	°C
SOTE as Function of Depth =	6.56%	% per m depth	Standard Pressure =	1.014	bar
AOTE/SOTE =	0.33		Atmospheric Pressure =	1.014	bar
Press. Drop across Diffuser =	0.030	bar	Air Density at STP =	1.200	kg/m³
(from mfr/vendor)			O_2 Content in Air =	0.2770	kg/m³

Figure 54. User Inputs for Oxygen/Air/Blower Calculations – S.I. Version

8. Oxygen/Air Requirement and Blower Calculations (for the Aeration Tank)

i) INPUTS (Values of "Rule of Thumb" Constants for the Calculations - See info at right)

O_2 needed per kg BOD =	1.50	lb O_2/lb BOD	Depth of Diffusers =	14.5	ft
O_2 needed per kg NH_4-N =	4.57	lb O_2/lb NH_4-N	Standard Temperature =	68	°F
SOTE as Function of Depth =	2.00%	% per ft depth	Standard Pressure =	14.7	psi
AOTE/SOTE =	0.33		Atmospheric Pressure =	14.7	psi
Press. Drop across Diffuser =	12.0	in W.C.	Air Density at STP =	0.075	lbm/SCF
(from mfr/vendor)			O_2 Content in Air =	0.0173	lbm/SCF

Figure 55. User Inputs for Oxygen/Air/Blower Calculations
U.S. Version

The results of the oxygen/air requirement and blower calculations are shown for S.I units in **Figure 56** and for U.S. units in **Figure 57**.

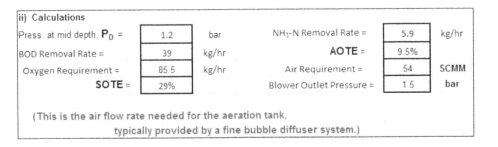

Figure 56. Oxygen/Air Requirement/Blower Calculations – S.I. Version

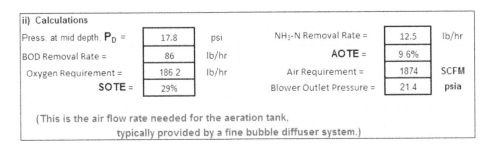

Figure 57. Oxygen/Air Requirement/Blower Calculations – U.S. Version

The calculations in this section are rather straightforward. The **pressure at mid depth** is the (diffuser depth/2) times the specific weight of water with unit conversions as needed. The **BOD removal rate** is the ww flow rate times the (influent BOD minus the target effluent BOD) with unit conversions as needed. The **NH$_3$-N removal rate** is the ww flow rate times (the influent TKN minus the target effluent NH$_3$-N). Again, unit conversions are needed. The **rate of oxygen requirement** in lb/hr or kg/hr is calculated as the **BOD removal rate** times the mass of O$_2$ needed per mass of BOD removed plus the **NH$_3$-N removal rate** times the mass of O$_2$ needed per mass of NH$_3$-N removed. The **SOTE** (standard oxygen transfer efficiency) is calculated as the diffuser depth times the specified % per unit depth for SOTE. The **AOTE** (actual oxygen transfer efficiency) is calculated as **SOTE(AOTE/SOTE)**. The **Air Requirement** is calculated (in SCMM) as [(**oxygen requirement** in kg/hr/**AOTE**)/**O$_2$ content in air** in kg/m^3]/60. To calculate the **Air requirement** in SCFM, the **oxygen requirement** will be in lbm/hr, and the **O$_2$ content in air** will be in lbm/SCF.

104

Calculation of Alkalinity Requirements are illustrated in **Figure 58** (S.I. units) and **Figure 59** (U.S. units). The equations used for those calculations are as follow:

Alkalinity used for Nitrification = 7.14(NOx)   (mg/L as $CaCO_3$)

Alk. Conc. needed = Alk. used for Nitrif. + Target Effl. Alk. - Alk$_o$
(mg/L as $CaCO_3$)

Alk. Flow needed = Q$_o$ (Alk. Conc. needed)/1000   (kg/day) - S.I. units

Alk. Flow needed = Q$_o$ (Alk. Conc. needed)*8.34   (lb/day) - U.S. units

Sodium Bicarbonate Flow needed = (Alk. Flow needed)(Equiv Wt. of $NaHCO_3$)/(Equiv. Wt. of $CaCO_3$)

9. Calculate Alkalinity Requirement					
Input: Target Effluent Alkalinity =	80	mg/L as $CaCO_3$			
Constants needed for Calculations:					
Equiv Wt. of $CaCO_3$ =	50	g/equiv.	Equiv Wt. of $NaHCO_3$ =	84	g/equiv.
Alkalinity used for Nitrification =	7.14	g $CaCO_3$/g NH_3-N			
Calculations					
Alk. used for nitrification =	196.9	mg/L as $CaCO_3$			
Alk. Conc. needed =	136.9	mg/L as $CaCO_3$	Alk. Flow needed =	1037	kg/day as $CaCO_3$
Sodium bicarbonate needed per day to maintain alkalinity =			1,742	kg/day $NaHCO_3$	

Figure 58. Calculation of Alkalinity Requirement – S.I. Version

9. Calculate Alkalinity Requirement

Input: Target Effluent Alkalinity = [80] mg/L as $CaCO_3$

Constants needed for Calculations:

Equiv Wt. of $CaCO_3$ = [50] g/equiv.   Equiv Wt. of $NaHCO_3$ = [84] g/equiv.

Alkalinity used for Nitrification = [7.14] g $CaCO_3$/g NH_3-N

Calculations

Alk. used for nitrification = [192.3] mg/L as $CaCO_3$
Alkalinity needed = [132.3] mg/L as $CaCO_3$ Alkalinity needed = [1656] lb/day as $CaCO_3$

Sodium bicarbonate needed per day to maintain alkalinity = [2,782] lb/day $NaHCO_3$

Figure 59. Calculation of Alkalinity Requirement – U.S. Version

Process Design Calculations for Pre-Anoxic Denitrification

Denitrification Background

In order to carry out denitrification of a wastewater flow (removal of the nitrogen from the wastewater), it is necessary to first nitrify the wastewater, that is, convert the ammonia nitrogen typically present in the influent wastewater to nitrate. The nitrification reactions require an aerobic environment and the denitrification reaction require an anoxic environment (the absence of oxygen). The anoxic denitrification reactor may be either before the BOD removal/nitrification reactor (called pre-anoxic denitrification) or after the BOD removal/nitrification reactor (called post-anoxic denitrification). Only the pre-anoxic option will be discussed here. A flow diagram for an MBR process with pre-anoxic denitrification is shown in **Figure 60** below.

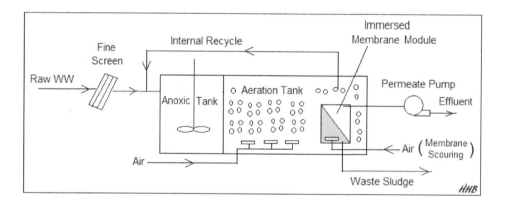

Figure 60. Flow Diagram for MBR Process with Pre-Anoxic Denitrification

In a pre-anoxic denitrification process, the BOD in the influent wastewater is used as the carbon source for denitrification. In this process, however the influent wastewater entering the pre-anoxic reactor still has ammonia nitrogen present rather than the nitrate nitrogen needed for denitrification.

A recycle flow from the aeration tank (identified as Internal Recycle in **Figure 60**) is used to send nitrate nitrogen to the denitrification reactor.

Process Design for Pre-Anoxic Denitrification

These process design calculations will be for pre-anoxic denitrification basins to go with the BOD removal/nitrification aeration tanks that were sized in the previous section. The wastewater parameters/characteristics and biological kinetic coefficients used above will also be used here. Additional user inputs needed for process design of the pre-anoxic basins are shown in **Figure 61**. The values in Figure 61 are for calculations in S.I. units. The U.S. version would have the freeboard specified as 1.5 ft, the liquid depth in the tank specified as 15 ft, and the mixing energy for the anoxic reactor would be 0.38 hp/10^3 ft^3. As noted at the bottom of Figure 61, a preliminary estimate for the anoxic detention time is needed for use in a later iterative calculation to zero in on its value.

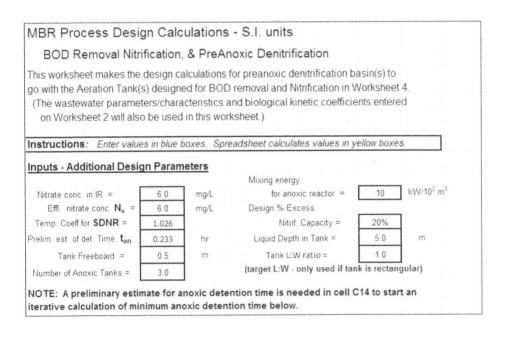

Figure 61. Additional User Inputs for Pre-Anoxic Denitrification Process Design - S.I. Version

Results of the first set of process design calculations are shown in **Figure 62**. The items calculated are the active biomass concentration, X_b, the internal recycle ratio, **IR**, the recycle flow rate to the anoxic tank, the NOx rate to the anoxic tank, the anoxic tank volume based on the estimated detention time, and the **F/M** ratio. The equations used for these calculations are as follows:

active biomass conc: $X_b = [\,Q(SRT)/V\,][\,Y(S_o - S)/(1 + k_d(SRT))\,]$

Internal Recycle Ratio: $IR = (NO_x/N_e) - 1.0$

Recycle Flow rate to anoxic tank: $Q_{anox} = IR(Q)$

NOx feed rate = $(Q_{anox})(NOX_{IR})$ $F/M = QS_o/XV$

Calculations - Preanoxic basin design for denitrification

1. Calculate active biomass concentration and IR ratio

Active biomass conc., X_b = | 6,278 | mg/L   IR ratio, **IR** = | 3.5 |

2. Calculate feed rate of NO_4-N to anoxic tank and anoxic tank volume

Recycle flow rate to anoxic tank = | 19,828 | m³/d NO_x rate to anoxic tank = | 118,969 | g/day
(flow containing nitrate)   Anoxic tank volume, V_{an} = | 55 | m³
                                                  (based on detention time value in cell C14)

3. Calculate the F/M ratio

Anoxic tank F/M ratio = | 2.868 | g BOD/d/g MLVSS

Figure 62. First Set of Calculations for Pre-Anoxic Denitrification Process Design -S.I.

The next set of calculations will calculate the value of the **SDNR** (specific denitrification rate) and make use of that **SDNR** value to calculate the required anoxic tank volume and detention time. The **SDNR** has units of g NO_3-N/day/g biomass. It is the rate at which nitrate can be denitrified (removed) in grams per day per gram of biomass.

Equations for SDNR as a function of F/M and rbCOD/bCOD
1. For rbCOD/bCOD = 0.10: SDNR = $-0.0761*(F/M)^2 + 0.263*(F/M) + 0.00636$
2. For rbCOD/bCOD = 0.20: SDNR = $-0.0674*(F/M)^2 + 0.270*(F/M) + 0.00385$
3. For rbCOD/bCOD = 0.30: SDNR = $-0.0608*(F/M)^2 + 0.278*(F/M) + 0.00149$
4. For rbCOD/bCOD = 0.40: SDNR = $-0.0591*(F/M)^2 + 0.289*(F/M) + 0.000536$
5. For rbCOD/bCOD = 0.50: SDNR = $-0.0558*(F/M)^2 + 0.300*(F/M) + 0.00268$

Figure 63. Equations for SDNR as a Function of F/M and rbCOD/bCOD

The calculated value for **SDNR** is based on Figure 8-23 in Metcalf & Eddy, *Wastewater Engineering, Treatment and Reuse*, 4th Ed. (Ref #1 below). Figure 8-23 is a set of graphs that give SDNR as a function of **F/M** and **rbCOD/bCOD**. Values were read from these graphs and used to derive the set of equations for **SDNR** as a function of **F/M** and **rbCOD/bCOD**. Those derived equations are shown in **Figure 63** on the previous page.

NOTE: The equations in **Figure 63** were derived from sets of values read from the graphs in Figure 8-23 in Metcalf & Eddy, Inc, (Revised by Tchobanoglous, G, Burton, F.L., Stensel, H.D.), *Wastewater Engineering, Treatment and Reuse,* 4th Ed., New York, NY, 2003.

Figure 64 shows the results of the calculation of the **SDNR** and required anoxic tank volume and detention time. Note that an iterative calculation is needed as described in blue at the bottom of **Figure 64**, to calculate the required Anoxic Tank Volume and Detention Time.

4. Calculate the SDNR

rbCOD/bCOD ratio =	0.321		Next lower rbCOD/bCOD value from table:	0.3
			Next higher rbCOD/bCOD value from table:	0.4

SDNR for rbCOD/bCOD =	0.3	is equal to:	0.3018
SDNR for rbCOD/bCOD =	0.4	is equal to:	0.3247

SDNR for rbCOD/bCOD = 0.321 is equal to: 0.3067 g NO_3-N/day/g biomass (at 20°C)

at T_{ww}, SDNR = 0.250 g NO_3-N/day/g biomass (at T_{ww})

5. Calculate the Anoxic Tank Volume, Dimensions and Detention Time
(User Input needed in Blue Cells)

NO_3-N reduction capacity = 142,683 g/d   % Excess Nitrif. Capacity: 19.9%
Difference between Design and Calculated % Excess Nitrification Capacity = 0.001

Goal Seek Result:
Anoxic Tank min. Vol., V_{an} = 91 m³   min. Anox. Det. Time, t_{an} = 0.38 hr

NOTE: This is an iterative solution. You must use Excel's "Goal Seek" to find the V_{an} v and t_{an} values as follows: Place the cursor on cell H51 and click on "goal seek" (in the "tools" menu of older versions and under "Data What If Analysis" in newer versions of Excel). Enter values to "Set cell" H51 "To value:" 0, "By changing cell:" C14, and click on "OK". The calculated values of V_{an} and t_{an} will be in cells C53 and H53. Cell H51 should be zero if the process worked properly. A preliminary estimate for t_{an} is needed in cell C14 in order for the iterative solution to work.

Figure 64. Calculation of **SDNR** and required Anoxic Tank Volume and Detention Time

The calculations for the values shown in **Figure 64** proceed as follows. **rbCODo** was a user specified value and **bCODo** was calculated as part of the BOD removal/nitrification process design calculations, so the ratio of those two can readily be calculated. Then If statements are used to find the two values of **rbCOD/bCOD** from the table in **Figure 65** that bracket the calculated value. The VLOOKUP function is then used to populate the yellow cells in **Figure 65** with the coefficients for the equations giving **SDNR** as a function of **F/M** at those two values of **rbCOD/bCOD**. Those coefficients are used to calculate the **SDNR** at each of the bracketing values of **rbCOD/bCOD** and interpolation is used to calculate the **SDNR** at 20°C for the **rbCOD/bCOD** value for this system.

Coefficients for SDNR Equation			
rbCOD/bCOD	Coeff. of F/M^2	Coeff. of F/M	Constant
0.1	-0.0761	0.2625	0.0064
0.2	-0.0674	0.2702	0.0039
0.3	-0.0608	0.2784	0.0015
0.4	-0.0591	0.2892	0.0005
0.5	-0.0558	0.2996	0.0027
0.20	-0.0674	0.2702	0.0039
0.30	-0.0608	0.2784	0.0015

Figure 65. Table of Coefficients for **SDNR** Equation

The denitrification capacity is calculated as: $V_{anox}(SDNR)(X_b)$

The % Excess Denitrification Capacity is then calculated using the denitrification capacity and the previously calculated NOx rate to the anoxic tank. Excel's Goal Seek function is then used to set the % Excel Denitrification Capacity equal to the User entered value by changing the anoxic detention time. This results in a minimum required anoxic tank volume and detention time.

Figure 66 shows the remaining calculated values for the pre-anoxic denitrification system. This includes the anoxic tank width and length, the reduced oxygen/air requirement due to the nitrate reduction oxygen credit, the reduced alkalinity requirement due to the alkalinity produced by denitrification, the anoxic tank mixing power needed, and the sludge wasting rate.

The oxygen credit for nitrate reduction is calculated as:
2.86 $Q_o(NO_x - N_e)$

The alkalinity produced by nitrate reduction is calculated as:
3.57(NO$_x$ - N$_e$)

The anoxic tank mixing power required is calculated as the anoxic tank volume times the user specified value for mixing power per unit volume needed.

The Sludge Wasting Rate is calculated as:
Q$_w$ = [(V*MLSS/SRT*TSS_W)

Min. Vol. per tank, V$_{tank}$ =	30	m^3	Click on green box and then on arrow to Select Tank Shape:	rectangular	
Calculated Tank Width =	2.5	m	Actual Tank Width:	2.5	m
Calculated Tank Length =	2.5	m	Actual Tank Length:	2.5	m
Tank Wall Height =	5.5	m	Actual Tank liquid Volume = (Pre-Anoxic Tank Volume)	31.3	m^3
Anoxic Det'n time, τ =	0.4	hr			

6. Recalculate the oxygen/air requirement due to the nitrate reduction oxygen credit

Oxygen credit =	340	kg/day	=	14.2	kg/hr	
O$_2$ Utilization Rate =	71.3	kg/hr	Req'd air flow rate, SCMM =	45.0	SCMM	
			Blower Outlet Pressure =	1.5	bar	

7. Recalculate Alkalinity Requirement

Alkalinity produced =	74.8	mg/L as CaCO$_3$			
Alkalinity needed =	57.6	mg/L as CaCO$_3$	Alkalinity needed =	327	kg/day as CaCO$_3$
Sodium bicarbonate needed per day to maintain alkalinity =			550	kg/day NaHCO$_3$	

8. Calculate Anoxic tank mixing power needed

Anoxic Tank Mixing Power =	0.9	kW

9. Calculate Sludge Wasting Rate

Waste A.S. Rate, Q$_w$ =	95	m^3/d

Figure 66. Calculations for Pre-Anoxic Denitrification System

References

1. Ardern, E. and Lockett, W.T., "Experiments on the Oxidation of Sewage Without the Aid of Filters," *Journal of the Society of Chemical Industry,* Vol 33, No 10, pp 523 - 539, 1914 (presented at Manchester Section meeting on April 3, 1914).

2. Metcalf & Eddy, Inc, (revised by Tchobanoglous, G, Burton, F.L., Stensel, H.D., *Wastewater Engineering Treatment and Reuse,* 4th Edition, New York, NY, 2003.

3. Bengtson, Harlan H., *A Spreadsheet to Calculate Oxygen Requirement Activated Sludge Process,* An online informational article at: www.EngineeringExcelSpreadsheets.com.

4. Bengtson, Harlan H., *Activated Sludge Calculations with Excel,* an online self-study course for Professional Engineer pdh credit.

5. Odegaard, Hallvard, "Compact Wastewater Treatment with MBBR." DSD International Conference Hong Kong, 12. 11-14-2014.

6. Odegaard, H., "The Moving Bed Biofilm Reactor," in Igasrashi, T, Watanabe, Y., Asano, T. and Tambo, N., Water Environmental Engineering and Reuse of Water, Hokkaido Press 1999, p 250-305.

7. Steichen & Phillips, H., M, Black & Veach, Process and Practical Design Considerations for IFAS and MBBR Technologies, Headworks International Presentation, 03/18/2010

8. Rusten, B and Paulrud, B., *Improved nutrient Removal with Biofilm Reactors*, Aquateam - Norwegian Water Technology Center, Oslo, Norway.

9. McQuarrie, J.P. and Boltz, J.P., *Moving Bed Biofilm Reactor Technology: Process Applications, Design and Performance*, Water Environment Research, Vol 83, No. 6. June 2011.

10. Bengtson, Harlan H., "MBBR Wastewater Treatment Design Spreadsheet." An online article at www.EngineeringExcelSspreadsheets.com

11. Judd, Simon, "The MBR Book, Principles and Applications of Membrane Bioreactors in Water and Wastewater Treatment,", 2nd Ed, Elesvier.

12. Zaerpour, Masoud, "Design, Cost & Benefit Analysis of a Membrane Bioreactor," M.S. Thesis, Department of Environmental and Geomatic Engineering, Potecnico di Milan, Academic Year, 2013-2014.

13. Yamamoto, K, Hiasa, H, Talat, M, Matsuo, T., "Direct Solid Liquid Separation Using Hollow Fiber Membranes in Activated Sludge Aeration Tank," *Water Science and Technology,* 21, 43 – 45.

Made in the USA
Middletown, DE
26 September 2023

39468135R00066